The Binocular Stargazer

A beginner's guide to exploring the sky

Leslie C. Peltier

Formerly published as
Leslie Peltier's Guide to the Stars

FROM THE PUBLISHERS OF ASTRONOMY® MAGAZINE

 Published by Kalmbach Publishing Co., 21027 Crossroads Circle, P. O. Box 1612, Waukesha, WI 53187.

Publisher's Cataloging-in-Publication Data
Peltier, Leslie C.
The binocular stargazer : a beginner's guide to exploring the sky / Leslie C. Peltier.
p. cm.
Includes bibliographical references and index.
Originally published as Leslie Peltier's guide to the stars
ISBN 0-913135-25-9

1. Astronomy—Amateurs' manuals. 2. Astronomy—Observers' manuals. I. Peltier, Leslie C. Guide *to the stars*. II. Title.

QB63.P45 1995 523
QBI94-21110

Contents

Foreword

This book is for those who would like to know the stars. To make this acquaintance, the only requirements are reasonably good eyesight, an inquiring mind, and a location from which you can see a fair expanse of sky at night.

Most emphatically, you do not need a telescope in order to become familiar with the stars and to explore the depths of space. Indeed, for the beginner who stands at the threshold of the night sky, it is far better that he make his first steps with no optical equipment other than binoculars. He who starts his wandering journey through the skies encumbered with a telescope seldom comes to really know the stars and constellations. All too soon he becomes enthralled by the splendor of Saturn's rings, by the craters and mountains of the Moon, or by the misty swirls of the Orion Nebula; he may be captivated by the circling moons of Jupiter or entangled in a devious maze of double stars, and thus he may fail to read the thrilling early pages of the sky's wide-open book. He will never come to feel the magic of a night that sparkles with a friendly company of stars that he can call by name. Nor can he ever file away in the storehouse of his mind the memory of his first sighting of some legendary figure rising in the east, a memory to be recalled and savored time and time again on other starlight nights all throughout his life.

—Leslie C. Peltier

"Five thousand years ago a young herdsman crouched beside his watch fire and saw the constellations rising star by star above the ancient hills of Nineveh.

Five hundred years ago an Indian lad, awakened in the dead of night by the mournful hooting of a great horned owl, peered out his tepee door and saw the flashing fire of the Mighty Hunter's Dog Star through the interlacing branches of the somber pines.

Fifty years ago a farm boy, on a perfumed night in May, found a blue-white guiding star beckoning to him from the dark sky just above the orchard trees.

Though widely separated by years, by miles, and by cultures, all three boys were bound by the common tie of starlight, for the same bespangled skies had looked down on storied hills, on trackless forest, and on the little farm. Each boy had felt, in his way, that vague but certain quickening that comes with the sighting of a familiar friendly star."

Although these words were written by my father the farm boy, almost twenty-five years ago, I know it is his wish that through the pages of this book others might be guided to the simple pleasures of an acquaintance with the stars.

—Dr. Gordon J. Peltier

Introduction

Our big Packard twisted into the driveway of just another Midwestern farm house, squatting like all of them on the glacial-drifted Ohio farmland. A couple of cows grazed in the yard, a Model T sagged behind the house, and a board gave passage over a muddy spot in the yard.

The plank-assisted path led into the nearby pasture, and in that pasture was a marvelous sight—a ten-by-ten foot frame shed with a dome on top. We had reached, for us at any rate, one of the great shrines of the world—the amateur observatory of Leslie C. Peltier. And it was sort of symbolic that the town nearby was called Delphos, the ancient source of revelation.

A slim, quiet man hardly older than I, greeted us. A vibrant sister Dorothy made the smiling welcome, and, as was customary in the Midwest, we entered through the kitchen door. None of us had seen the other before but we embraced a mutual hobby—we observed variable stars.

This was my introduction to Leslie Peltier, who at the age of 32 was established as America's leading amateur astronomer. Eventually he was to become even more luminous than he was that spring day in 1932. But already he had racked up two comets or so, three novae, and more variable star observations than anyone else in the country.

The country supper food was good but I do not remember a dish. It was clear outside, and we would be able to observe that night and match our newly acquired observational skills against the Master of them all. We would take turns estimating the exact brightness of perhaps four dozen variable stars.

Peltier joined the American Association of Variable Star Observers in 1918, so on that night in 1932 he had fourteen years of experience to my one year. We trudged out that night under skies that Delphos seldom has today. Up went a ladder, up went Peltier, and down was handed the dome slit cover. Farm boy as he was, there was no need for a fancy mechanism to take a cover off.

Inside sat a telescope, gleaming wood, that was already notable. The whim of a Princeton astronomer to loan him the scope had shoved Peltier onto the fast track to astronomical fame, a position he maintained with ease and distinguished style for the next half century.

At Princeton it had captured three comets; Peltier had added three more. The names were deeply carved on its mahogany tube. It was a 6-inch, which was large for the AAVSO at that time. As we admired it Peltier noted that if it had been a non-comet telescope he probably would never have gone on to find his but would have stuck only to variables.

The observing session started as soon as it was dark. We took turns estimating the brightnesses of variable after variable. Leslie found all the fields to speed things up. His scope had no finder; he merely looked along the tube, nudged his scope a trifle, then lo and behold, the variable field would be in the eyepiece. He never was wrong! We had heard of this talent of his, but still it dazzled us.

Later after a ham and egg breakfast, and coffee with grounds, we compared the results of the observing session. I was no more than one-tenth of a magnitude off Peltier. Talk of mead and honey! To me it meant I had graduated, I had gotten around the last buoy in the race, my doubts about myself were quenched by a Peltier fire hose. A great adventure could now begin.

But while I was instantly grateful to Peltier, I did not realize that I was to be only one of many whom Leslie conducted quietly into what may be called significant activity in astronomy, activity that was a contribution to the world I played in.

I didn't dream that Leslie would send dozens of young people into professional astronomy. Some of them now man giant telescopes. More have found that serenity that marked Peltier and have learned that the stars are better than a witch doctor. First it was his example, then it was his book, *Starlight Nights,* which had the widest impact of any astronomy book since Garret P. Serviss wrote at the turn of the century. Perhaps time will show that Peltier did even more good.

Dr. Lewis Epstein once told the Astronomical Society of the Pacific that the chief job of the amateur astronomer has always been to insure the continuation of the attitudes that have made astronomy great. This is more important than making variable star or meteor observations, valuable as they are.

Peltier that night did a good deal for my attitudes, and in return we invited him to continue with us to Maryland and the AAVSO spring meeting. Although he turned shy instantly and mumbled about farm chores, we finally persuaded him to come along.

The small convention was ecstatic when Peltier thus came to his first convention. He liked it, too. It was his first chance to talk to observers he knew only by mail. All the way home to Delphos he kept smiling his small bright smile.

In fact, he was so pleased he then assaulted the heavens with redoubled efforts and in a few months—there was another comet discovered.

—*Walter Scott Houston*

1
Trail's Beginning

A guidebook should start at the beginning of the trail. This is especially true when the pathway leads through unfamiliar territory. Right here, then, at the very outset of a starwalk that will wind all the way across the skies and through all the seasons of the year, I will assume you have never passed this way before. You don't know the name or location of a single star. For you, the comings and goings of the planets are a total mystery. The Moon's round face is but a spotty globe, and you have never seen a sunspot or watched the variations in the brightness of a star.

To you, the Sun, the stars, the planets may have always appeared changeless when contrasted with a warbler in a springtime treetop. Your view of the skies seemed just the same from day to day and night to night. That was an illusion. You were really watching an animated scene, though you saw it from so great a distance that no movement caught your eye.

In coming chapters I will direct your eyes and binoculars to a number of locations where you'll find abundant proof that the sky is a busy place indeed, both day and night. You'll see sunspots larger than Earth. Sometimes you'll watch them grow from day to day as they move across the Sun. At other times you'll see them fade and disappear or even circumnavigate that turning, gaseous globe. By night you'll watch a sunrise two weeks long as it slowly moves across the Moon, while, in season, you will time the traffic of four other moons about a distant neighbor planet. On other nights, you'll meet a class of restless stars that will fade or brighten by a thousandfold.

Yes, your "changeless" skies are filled with action, with mighty movements that dwarf anything we know here on Earth. And you can spot these changes in the skies and turn your binoculars on them with ease when you become familiar with their habitat, the stars and constellations. We learn best by doing. The study of geology becomes more meaningful

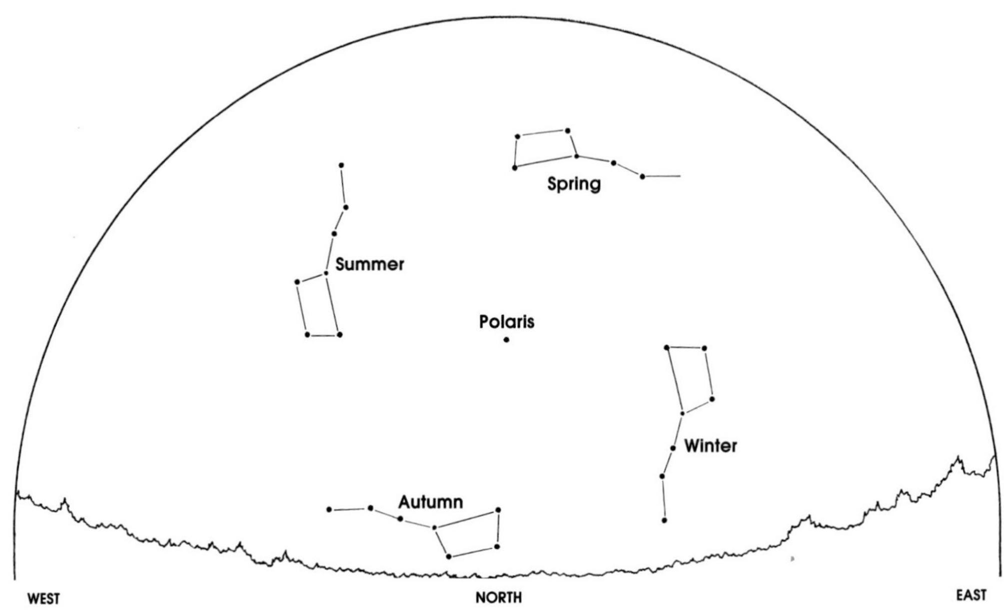

Seasonal Motion of the Big Dipper
Circling around the star Polaris, the Big Dipper moves across the sky from season to season. Observing at about 9 p.m., the constellation is east of true north on January 15; overhead on April 15; west of true north on July 15; and low on the northern horizon on October 15.

when we can hike to mountain, mine, or quarry and see the rocks in place. Botany takes on new life and interest when we plant a seed and watch it sprout and gradually develop roots and leaves and flowers. We learn to identify birds more quickly and effectively by stalking them at their activities in tree or shrub or meadow. In this same way, by involvement, the starry skies will cease to be a trackless maze. They will seem to come down to meet you and become a part of your own neighborhood when you go out on your first stargazing night and quickly find a certain star or star group.

THE BIG DIPPER

On any clear night of the year—preferably when there is no bright moonlight—go out to some location that is fairly free of trees, lights, and buildings and face the northern sky. Somewhere in this half of the sky dome, you are certain to see a group of seven conspicuous stars arranged in the shape of a dipper. Appropriately enough, this asterism is usually known as the Big Dipper, and the drawing at the top of this page shows it just as you will see it in the northern sky at a convenient evening hour near the middle of each of the four seasons of the year.

If your first stargazing session is on a clear winter night—say around

January 15 at about 9 p.m.—you will find the Dipper somewhat to the right of true north and apparently standing upright on its handle, as I have shown it here. Three months later—on April 15 at the same mid-evening hour—the Dipper can be found almost directly overhead. In July it will appear in the west, and on October 15 you will find it low in the north, where the autumn mists of the horizon may make it difficult to see all seven stars.

You can see that the Big Dipper seems to make a counterclockwise circle in the northern sky in one year's time. Oddly enough, it also makes this same circle in just one day! This is something you can easily check for yourself. If you look for the Dipper an hour or two after your 9 p.m. sighting on January 15, you will find those seven stars noticeably higher in the sky. If you delay this second sighting a full six hours—to 3 a.m. on the morning of January 16—you will find the Big Dipper high above your head in the April 15 position. Thus, the apparent daily movement of the stars from 9 p.m. to 3 a.m., or one-fourth of a day, equals the apparent seasonal movement from January 15 to April 15, or one-fourth of a year.

In watching the Big Dipper's changing positions you may already have noticed that it seems to be revolving slowly about a central point in the sky marked by a star of about the same brightness as the Dipper stars. This star is called Polaris, the North Star, for it is close beside the north pole of the sky. Note, too, as you continue to watch this circling Dipper, that no matter at what hour of the night or at what season of the year you look, the two end stars in the Dipper's bowl point almost directly to Polaris. For this reason, these two stars are called the Pointers.

This circling of the Big Dipper about Polaris clearly demonstrates the apparent movement of the stars. I use the word *apparent* here, for the ceaseless circling you see is due not to any actual movement of the stars themselves, but to the effect produced by two separate motions of the Earth. The spinning of the Earth on its axis—from west to east once every twenty-four hours—not only gives us night and day but also creates the illusion that the stars of the Big Dipper are revolving around Polaris once each day.

Earth's other movement, the year-long journey it makes around the Sun, causes the slow seasonal shifting of the positions of the stars. We are going to meet the stars that lie before us in the east, and this forward, eastward movement constantly changes our position in respect to all the stars, causing each star to arrive at its same position in the sky almost four minutes earlier each night.

You can easily verify this if you watch some star rising in the eastern sky. If, when viewed from your fixed position, this star clears the rooftop of a building at exactly 10 p.m., it will emerge at 9:56 on the following night. These four-minute nightly gains add up to two hours each month and twenty-four hours in a year, so one year after your first sighting, that star will come over the same rooftop again at exactly 10 p.m.

In addition to this apparent east-to-west motion of the stars—which you can readily detect in half an hour of attentive watching—each star has a motion of its own. It may be at a speed of thousands of miles per minute, but the stars are so distant they seem fixed in their positions in the sky relative to each other. However, in the long view of time they are anything but fixed. They move relative to each other in space and, given enough time, in our sky. No stars show this proper motion, as it is called, more clearly than the stars of the Big Dipper.

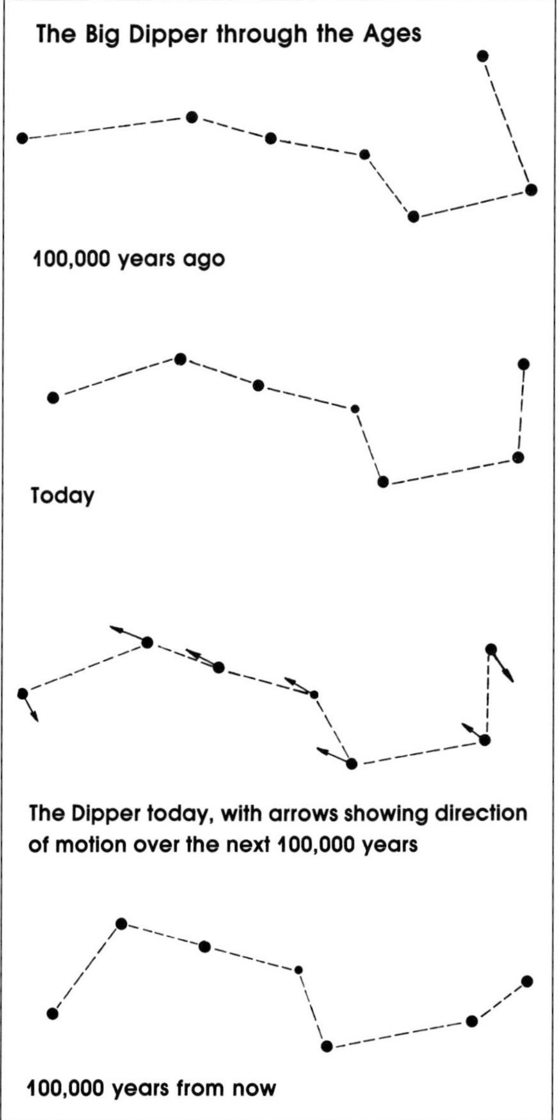

In just 100,000 years, you won't even recognize the Big Dipper.

The drawings on this page illustrate a chapter in the long life story of these seven stars. The upper sketch shows the Dipper as it appeared to early humans some 100,000 years ago. Next, it's shown as we see it today. Below that is today's Dipper, with arrows showing the direction and amount of motion of each star over the next 100,000 years. Note that two of these stars, the upper Pointer star and the star in the end of the handle, are moving in a direction nearly opposite that of the other five stars. This suggests that these five all belong to a moving star group with a common origin. In 100,000 years, as the bottom figure shows, the handle of the Dipper will appear badly bent and its bowl will flatten out until the Pointers will no longer serve to lead the eye to bright Polaris.

CONSTELLATIONS

Stargazers in the valleys of the Tigris and Euphrates Rivers probably devised our constellations some 5,000 years ago. Living outdoors as shepherds, farmers, desert nomads, and tribal warriors, they saw in the random groupings of the stars resemblance to familiar things. They peopled the skies with other herdsmen, hunters, and archers. They found in those early starfields their domesticated and wild animals such as dogs, bears, horses, and bulls. Later, when the Greeks learned of these sky pictures and stargazed themselves, the folklore and mythology of their country was also hung in the celestial gallery in the form of a king, a queen, and many of their fabled heroes.

The constellations, fanciful though they were in outline and design, served a worthy purpose. They gave each star a family name as well as a home territory. I can't imagine how anyone could possibly learn the stars if each one was just an individual with no close relatives or other family ties. Even the most accomplished stargazer can't recognize stars that appear by themselves. They must be in their proper setting, surrounded by their neighbor stars, to be identified. By segregating the stars into forty-eight constellations, each depicting some familiar bit of folklore, those early skywatchers made the first step toward bringing order to the stars.

The sky's eighty-eight constellations are listed in Appendix IV, together with their abbreviations and a space for you to record the date of your first sighting of them.

URSA MAJOR

The ancient weavers of the sky patterns placed the seven stars of the Big Dipper in the constellation Ursa Major, the Great Bear. As they saw it in the sky, the handle of the Dipper forms the bear's tail, while the bowl extends well to the middle of the creature's back. One faint star marks the bear's nose, while three pairs of stars outline the feet. Don't expect to see anything about this star group that in any way resembles a bear, for it simply isn't there. A modern attempt at rearranging all constellation outlines for greater realism shows the bear facing in the opposite direction with its nose marked by Alkaid, in the end of the Dipper's handle. But this bear is just as far-fetched as before, and the sky thereby loses some 5,000 years of history.

Fortunately, in our study of the skies it is not necessary to trace out any of these ancient patterns, for only in three or four cases is there any actual resemblance between the constellation figure and the earthly prototype for which it was named. It is more important to become familiar with each star group as you see it. For example, you should be able to quickly locate the Big Dipper at any time of night and any season of the year.

Then you can use its stars to guide your eyes and your binoculars to any feature of interest in the entire constellation. Furthermore, as you have already seen, the Big Dipper also obligingly points the way to still other stars and constellations, such as Polaris in the handle of the Little Dipper.

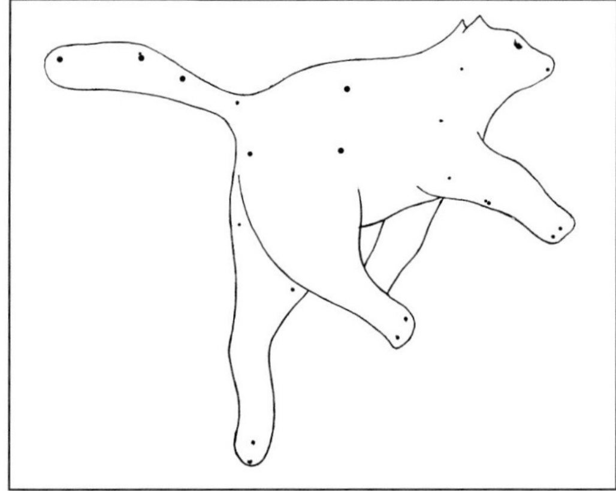

Ursa Major, the Great Bear, as the ancient stargazers saw it in the sky.

2

The Little Dipper and Polaris

To the nameless namers of the stars, the Little Dipper formed the constellation Ursa Minor, the Little Bear. Here as in the Great Bear, we find the figure of a dipper made up of seven stars, four of these in the bowl and three in the handle—although in this case, the handle curves up instead of down.

On ancient star charts, the tail of the Little Bear is as long as the creature's body, with its tip marked by a bright star, Polaris, which now stands near the north pole of the sky. Actually, the true celestial pole—the point exactly in line with the Earth's 8,000-mile-long axis—is two Full-Moon diameters distant from Polaris and lies on a line drawn from Polaris to the star Mizar in the handle of the Big Dipper. This means that Polaris, like all the other visible stars, circles around the true pole once every twenty-four hours. You can see this for yourself by taking a time-exposure photograph of the polar region some clear night. Your photo will show Polaris as a bright, circular arc whose center is the sky's true polar point.

Polaris may well become the first star you enter in your life-list of celestial objects (see Appendix V). It's a fine beginning, for it is easily the most useful star in all the night sky. For long ages, Polaris has served as man's guiding light in his nocturnal wanderings. The early Nordic sailors learned to steer their ships with this star as their navigating beacon. In later times, both pioneer and plainsman depended on the friendly reassurance of its gleam. Nineteenth-century poet William Cullen Bryant in his "Hymn to the North Star" pays this tribute to the star's importance:

On thy unaltering blaze,
The half-wrecked mariner, his compass lost,
Fixes his steady gaze,

And steers, undoubting, to the friendly coast;
And they who stray in perilous wastes, by night,
Are glad when thou dost shine to guide their footsteps right.

Today Polaris is still a star everyone should know, for whoever looks directly at it knows that he or she is facing north, that east is to the right, west to the left, and south to the back. This star's steadfast glow can guide the beginning skywatcher's early exploration of the skies.

Polaris is also an aid in determining your own latitude on Earth's surface, for the north pole of the sky is the same number of degrees above the northern horizon as you are north of Earth's equator. For example, if you live in New York City, Philadelphia, Pittsburgh, Indianapolis, or Denver, you are roughly 40 degrees north of the equator. For you, Polaris would be 40 degrees above the northern horizon, or a little less than halfway up the sky from the horizon to the zenith, the point directly overhead. The observational information in this book is oriented to stargazers living at this approximate latitude. As you travel north, Polaris appears to climb higher and higher in the sky. At Minneapolis, 45 degrees north latitude, the pole star is halfway up the sky. At the North Pole, the celestial pole is directly above you, at your zenith, with Polaris circling closely about it every twenty-four hours.

If you were standing at the North Pole you could only move in one direction—south. You are already as far north as you can go and every point on Earth is south of you. Rudyard Kipling was skirting scientific truth when he wrote:

For there is neither East nor West,
Border nor breed nor birth,
When two strong men stand face to face,
Though they come from the ends of the Earth.

His "two strong men" must have been standing on the poles, the ends of the Earth, for only at those points is there neither east nor west.

The configuration of the Little Dipper is not nearly as conspicuous as that of the Big Dipper. In fact, in moonlit or hazy skies, your unassisted eye may have some difficulty in locating all seven of its stars. Only Polaris (α) and Kochab (β) are as bright as the stars of the Big Dipper.

The brightness of stars is of particular interest to stargazers. In the pages to follow, I will often describe a star's brightness, measured by its magnitude. This is a measure of a star's relative brightness as we see it from Earth. For example, Sirius, the brightest of all the stars, appears so bright to us because it is the nearest of all the naked-eye stars visible from our latitude. On the other hand, Canopus, the next brightest star, though too far south for us to see, is some thirty times as distant as Sirius and owes its brilliance to its truly brilliant light output. If we could view these two stars from some point in space midway between them, Canopus would greatly outshine Sirius.

About 6,500 naked-eye stars lie scattered over the sky. In the second century, the Greek astronomer Claudius Ptolemy divided them into six "degrees" of brightness: from the very brightest, which he called 1st magnitude, down to the faintest he could see, which became 6th magnitude. He determined a light ratio of 2.5 between each magnitude; that is, a 2nd-magnitude star is two-and-a-half times fainter than a lst-magnitude star, and a 3rd-magnitude star two-and-a-half times fainter than a 2nd-magnitude, and so on.

Ptolemy's system of magnitudes is still used today. You'll find the precise magnitude of all bright stars in Appendix III. About twenty 1st-magnitude stars exist. There are about sixty 2nd-magnitude stars in the sky—three times the number of lst-magnitude stars. The same brightness ratio and threefold increase hold true for the other magnitudes, resulting in about 4,500 sixth-magnitude stars over the sky—each of which is 1/100 as bright as a lst-magnitude star.

The Little Dipper conveniently furnishes examples of each of these naked-eye magnitudes, except it contains no 1st-magnitude stars. Because these stars are all well up in the sky every clear night throughout the year, you can quickly refer to them to estimate the brightness of other stars. Thus Polaris, at magnitude 2.0, is a standard 2nd-magnitude star, while Kochab, at magnitude 2.1, is only slightly fainter.

The four stars of the Little Dipper's bowl represent four different magnitudes. Three of them, Beta (β), Gamma (γ) and Eta (η), can be used as standard 2nd-, 3rd-, and 5th-magnitude stars, while Zeta (ζ) is a rather faint 4th-magnitude object. And the faint star just below the bowl is a close example of a 6th-magnitude star.

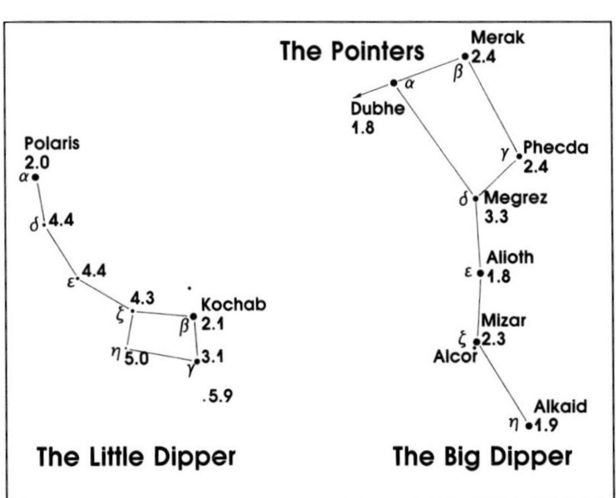

A close-up of the dippers with the proper names, Greek letter designations, and magnitudes of the brighter stars of both dippers.

Don't be discouraged if you have some difficulty tracing out all the stars of the Little Dipper. This isn't an easy figure to see with the naked eye, as it requires a dark, moonless night and a location free from interfering lights. But even under poor conditions, the Dipper stars and many others even fainter are well within the grasp of your binoculars.

Your eyes will always be the most precious and versatile instruments at your command. True wide-angle lenses, your eyes take in a vast expanse that reveals the entire picture you are looking for. As automatic zoom lenses, they hold in constant focus both the near at hand and the far away. But as you doubtless learned on your first field trip as a Scout or camper, your eyes could often use more power.

This same need for added strength is experienced by those who follow star trails, for within these constellations and scattered all about the sky there are celestial objects of interest and charm that lie beyond the reach of the unaided eye. Hidden there are sparkling star fields, glowing galaxies and nebulae, compact clusters of a million suns, and restless stars that change in brightness from one night to the next. You may see double stars and variable stars, roving fellow planets, circling satellites, and the ever-changing face of our Moon. You can look for streaking meteors and, more rarely, a passing comet or an exploding star. All of these are visible through binoculars.

These first two chapters have shown you how easy it can be to meet the stars and recognize them. You are now familiar with the naked-eye appearance of the two most important constellations in the sky: Ursa Major, with the seven bright stars of the Big Dipper, whose Pointer Stars then led you on to Polaris, and the Little Dipper in the constellation Ursa Minor. These two star groups can serve you well in further exploration, for they are year-round beacons that shine throughout the night to orient and guide you to other stars and constellations.

3

Augmented Eyes

Most modern binoculars are prism binoculars. They come in so many styles, sizes, powers, field widths, and prices that their respective specifications should be carefully examined before making a choice.

Prism binoculars have that name because each tube or barrel contains a set of glass prisms that intercept the incoming light and pass it through a series of reflections. The arrangement allows packing a longer focal length into the instrument without adding to the length of the barrels. This optical system was invented over a century ago by a French engineer named Porro, and the prisms he devised still bear his name. Some binoculars employ another system, in which straight barrels are used rather than the offset barrels of the porro prism models, resulting in a lighter, more compact instrument. This type is also more expensive.

Somewhere on the body of the binoculars you will usually find a series of figures that tell the power of the binoculars and the diameter of the objective or front lenses. Most often these figures will read 7x35 or 7x50. In either case, the 7x means the binocular magnifies an object seven times, while the 35 or 50 indicates an objective lens that is 35mm or 50mm in diameter. 7x binoculars are a great favorite of bird watchers—and with good reason. A power of seven is about the limit of magnification that you can hold steady in your hand. If binoculars are much

Darla Evans

Binoculars come in a wide variety of sizes and designs.

stronger than this, they magnify every pulse beat, nervous tremble, and gust of wind. Plenty of light passes through 35mm lenses under average conditions, but near twilight or in deep woods the 50mm objective works better—though it also adds to the weight, bulk, and certainly the expense of the instrument.

Powers of 10x and greater are justified for special users like wildlife watchers whose quarry, such as deer in the forest, cannot be closely approached—though here the use of a tripod or some form of steady mount will greatly help. If your stargazing studies require a larger image of the Moon or planets, then 10x is an advantage. Here, perhaps, you can devise a more permanent mounting in a suitable location and take the binoculars in after each use.

Prism Binoculars

Among the finer features of prism binoculars is the broad, sharp field of view they afford. This field dimension is often printed on the instrument, telling the width of the field (in feet) at a distance of 1,000 yards. Most 7x35 binoculars have fields ranging in width from 367 to 420 feet. These are normal fields. However, another class known as wide-field binoculars have fields of more than 500 feet. Originally designed for the sportsman, they permit a more inclusive view of the playing field and the race track. In constellation hunting, a large field of view is supreme because it allows a single field to hold the entire pattern of a star group or the flowing river of the Milky Way from shore to shore.

The diameter of the exit pupil of binoculars determines the amount of light received by your eye. The size of the exit pupil can be found by dividing the diameter of the objective by the power. In a 7x35 glass, the exit pupil is 5mm, which is ample in ordinary light, as it is larger than the 4mm diameter of the pupil of your eye. 7x50 binoculars have a 7.1mm exit pupil, which is better in poor light and allows still more leeway in case you can't hold the binoculars perfectly steady. If your binoculars have an exit pupil less than 4mm in diameter, you are not using your own eye to its full capacity.

Coated optics are an improvement older binoculars lack. They consist of a thin coating of magnesium fluoride, which reduces glare and reflections on internal surfaces. These coatings can be seen as reflected blue or amber colors when held at an angle to the light.

If you are buying binoculars, a great variety are available at prices ranging from $50 to $1,000. At the high end of this scale are the binoculars offered by the companies that make expensive cameras. There are, however, many mail order houses that carry good binoculars in the low-to-medium price range, say from $75 to $300, that will perform splendidly and bring just as much pleasure as the high-priced luxury models.

Before purchasing any binoculars, new or used, make a thorough quality test. Most important, check for image quality. Stars, for example, should be clear and sharply defined over the entire field of view. Both barrels of the instrument should be perfectly aligned so they form a single bright image with no overlapping. Be sure the barrels can be adjusted to the correct interpupillary separation of your eyes.

You can test the quality of the glass prisms in your binoculars by holding the instrument a few inches in front of your eyes and pointing it to the open sky. With the best quality binoculars, the exit pupils will show as two circular disks of light. With cheaper instruments you will see a square or oblong within the bright circle of each pupil.

If the power of the binoculars is not printed on them, you can determine it. Aim your binoculars at a brick wall or at the side of a frame house, then look through one barrel with one eye while your other eye surveys this same target outside the barrel. How many bricks or siding boards, as seen with the naked eye, does it take to fill the space of just one brick or board as seen through the glass? If seven are required, you have 7x binoculars.

The Big Dipper will help you estimate the angular width of your instrument's field of view, using 10 degrees as the distance across the top of the bowl, 7 degrees across the bottom, and 5 degrees between the Pointers.

Now, with your natural curiosity, patience, and perseverance, you are ready to greet the night skies of every season of the year. The more familiar you become with the stars, especially with their naked-eye appearances, the more readily you can train your binoculars on the objects hidden there.

As an introduction to these brighter stars and constellations, I have drawn a series of maps and charts that picture the skies as seen from our mid-northern latitudes. You have already met three trusty guides to lead the way: the Big Dipper, the Little Dipper, and Polaris.

4
Maps of the Sky

In an earlier chapter you became familiar with the figure of the Big Dipper. You saw how easy it was to locate the North Star, Polaris, in the Little Dipper merely by following with your eye the line projected by the two bright Pointer Stars. By this same simple method you can discover all the other stars and constellations. The skies have many other pointer stars and reference lines that will introduce you to neighbor stars or star groups, which in turn will point to still others, until before long you will be familiar with all the brighter stars and constellations that cross your skies throughout the year.

In the following series of sky maps and key charts you will find the stars and constellations pictured as they appear at the season of the year when they are in the best position for viewing. Because haze and sky-glow as well as trees and buildings sometimes obscure the horizon, the best observing time for most of us comes when the object of our search is on or near the meridian, an imaginary line that passes through the north and south celestial poles and our zenith. A star on the meridian is at its highest point in its daily travel.

Each sky map and key chart shows the sky as it appears on the semimonthly dates at the Standard Time hours listed. For daylight savings time add one hour, and for dates and hours other than those given, the appearance of the sky will be a close approximation to the nearest given date and evening hour. The sky maps on the right-hand pages show the nighttime appearance of the skies with some landscape features added for scale and realism. On the left-hand pages directly opposite the sky maps are the key charts. Along each margin are the compass directions covered by each chart. Here the stars are plotted exactly as they appear on the sky maps, but now the stars, constellations, and various features of interest have been labeled with their names. The Latin names of the constellations appear in larger type, while the proper names of many of the brighter stars and the Greek letter designations are given in smaller type. Also in the key charts many of the more prominent stars have been connected with solid lines in order to make a definite design.

Throughout the book I have given the English names of the Greek letters. The entire Greek alphabet follows:

Alpha	α	Iota	ι	Rho	ρ
Beta	β	Kappa	κ	Sigma	σ
Gamma	γ	Lambda	λ	Tau	τ
Delta	δ	Mu	μ	Upsilon	υ
Epsilon	ε	Nu	ν	Phi	φ
Zeta	ζ	Xi	ξ	Chi	χ
Eta	η	Omicron	ο	Psi	ψ
Theta	θ	Pi	π	Omega	ω

There are two sets of maps and charts for each season of the year. The first set for each season shows the half of the sky you see when you are facing north; the second set depicts what you see when you face due south. Inasmuch as the sky presents the appearance of an inverted bowl, it is impossible to show it on a flat page without some distortion around the edges. This shouldn't cause any problems, as most of our serious observing will be confined to the mid-regions with the least atmospheric interference, where distortion is minimal. The letter Z, high up on each key chart, marks the location of your zenith, the point in the sky directly above you, a point in the sky that's yours alone.

5
The Winter Stars

FACING NORTH

In the sky map on page 24, which covers the northern half of the sky, are two star figures that you have already identified. First you located the Big Dipper with its two Pointer Stars, which introduced you to the second figure, the Little Dipper and the star Polaris.

To the south and east of the Big Dipper's bowl you will find a number of fainter stars and these, together with the seven stars of the Dipper, make up the constellation Ursa Major, the Greater Bear. The best known star of this large constellation is Zeta, or Mizar, the middle star in the Dipper's handle. Even a casual glance at this star will show that it is not alone, for a 4th-magnitude star named Alcor travels through space right beside it, making a naked-eye double star. There are thousands of double stars scattered about the sky, but Mizar is more than doubly famous. Not only was it, with Alcor, probably the first known naked-eye double, but Mizar itself was the first double star discovered with a telescope. This happened in 1650 —some forty years after the invention of the telescope—which would seem to show that the early sky watchers were more concerned with the Moon and planets than they were with stars.

But there is even more to Mizar than meets the eye in any telescope. The brighter of the telescopic pair later proved to be a spectroscopic double star and, again, it was the first one ever found. Too close to be detected in a telescope, the periodic doubling of its spectral lines showed Mizar to have a faint companion star revolving about it every 20 days at a distance of about 36 million miles—less than half the distance between Earth and the Sun.

There are two main types of double stars: optical doubles and binary systems. Optical doubles are those in which two stars happen to be nearly in the same line-of-sight. This makes them appear to be side by side when, in reality, one of the pair is farther away from us than the other.

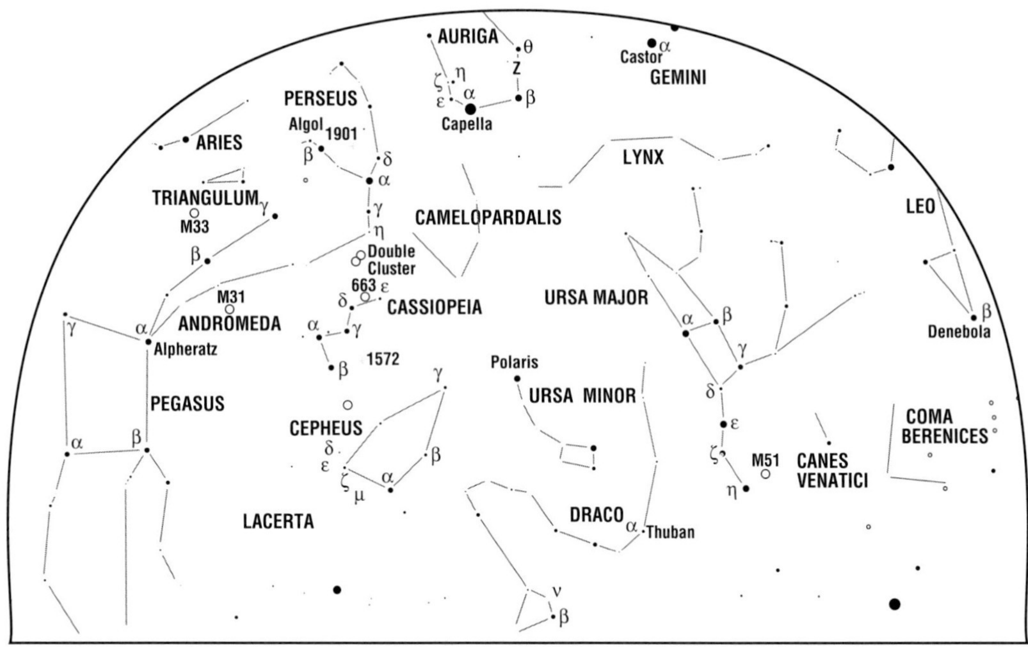

Constellations of the Northern Winter Sky
Andromeda Camelopardalis Cassiopeia Cepheus Leo Minor Ursa Major Ursa Minor

Binary stars, on the other hand, are true doubles, for they form a system of two stars revolving about their common center of gravity. As in the case of Mizar and its faint companion, binary stars are close to each other in the sky. Contrary to what you might expect, about one-third of all stars belong to binary or multiple star systems—all visible proof that gravitational attraction is a universal law.

Well to the west at this season of the year, the Milky Way sweeps upward from the horizon to the zenith. The lower part, however, is often contaminated by the haze and heavy atmosphere near the horizon and by the interference of trees and buildings, unless you can get out into open country away from such distractions. I realize that few are free to do this except on rare occasions, and so I will discuss the stars and constellations at the times and seasons when they are well up in the sky where they can be seen through our thinnest atmosphere.

In the western sector of this northern winter sky some 2nd-magnitude stars can be seen in the Milky Way, but they have already passed their prime position in the sky, and so I will describe them later when they are rising in the east as autumn stars. There is a dearth of lst-magnitude stars in the northern sky at this time of year. Only one such star, Capella, can be found; it is now so nearly overhead that it can be more comfortably

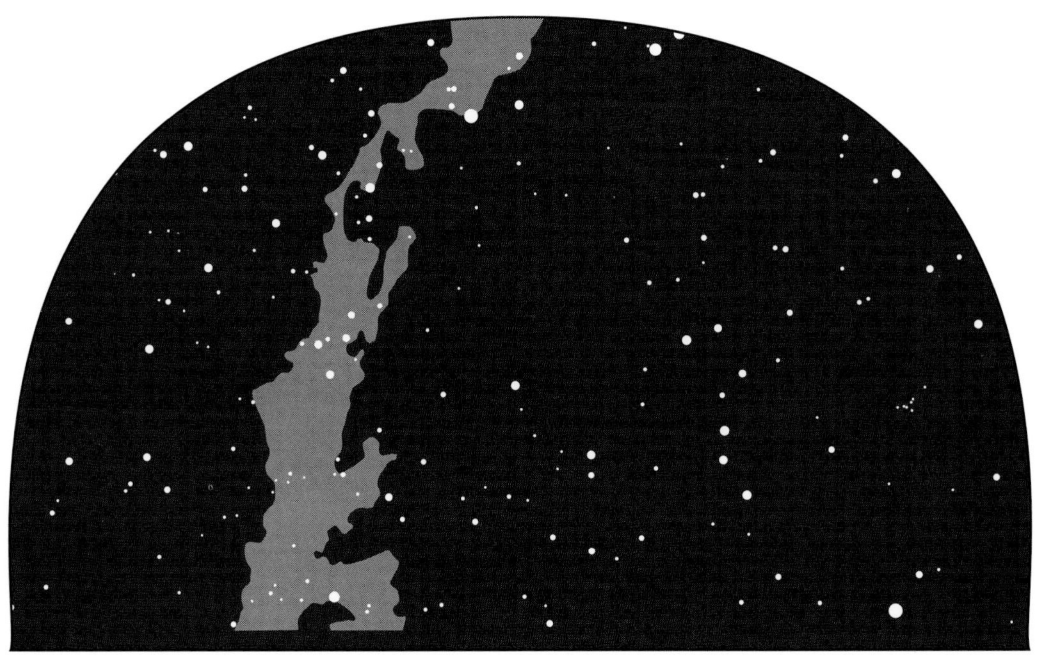

The Northern Winter Sky As Seen On:

December 15 at 12 a.m.	*January 15 at 10 p.m.*	*February 15 at 8 p.m.*
January 1 at 11 p.m.	*February 1 at 9 p.m.*	*March 1 at 7 p.m.*

observed when it will be climbing in the eastern sky in autumn. Perhaps this scarcity of northern stars is just as well right now, for as you turn and face the south the year's most brilliant skies are on display.

FACING SOUTH

If you have ever wondered what to do on long winter evenings, the southern skies provide the answer. Here, scattered about from east to west and from zenith to horizon, are no less than eight 1st magnitude stars—more than half the total number of these stars that you can see from mid-northern latitudes. Nowhere in the heavens will you find another such glittering array, for six of these bright stars are now grouped near the center of the winter stage.

ORION

The most brilliant of all constellations is Orion, the Hunter, whose bold figure now stands upright in the southern skies just west of the meridian. With the exception of the Big Dipper, Orion is the best known star group in the sky and rightly so, for it includes two stars of 1st magnitude, four of 2nd magnitude, as well as a host of fainter stars. In my earliest atlas of the skies, Orion is shown as a giant who holds an uplifted club in his right

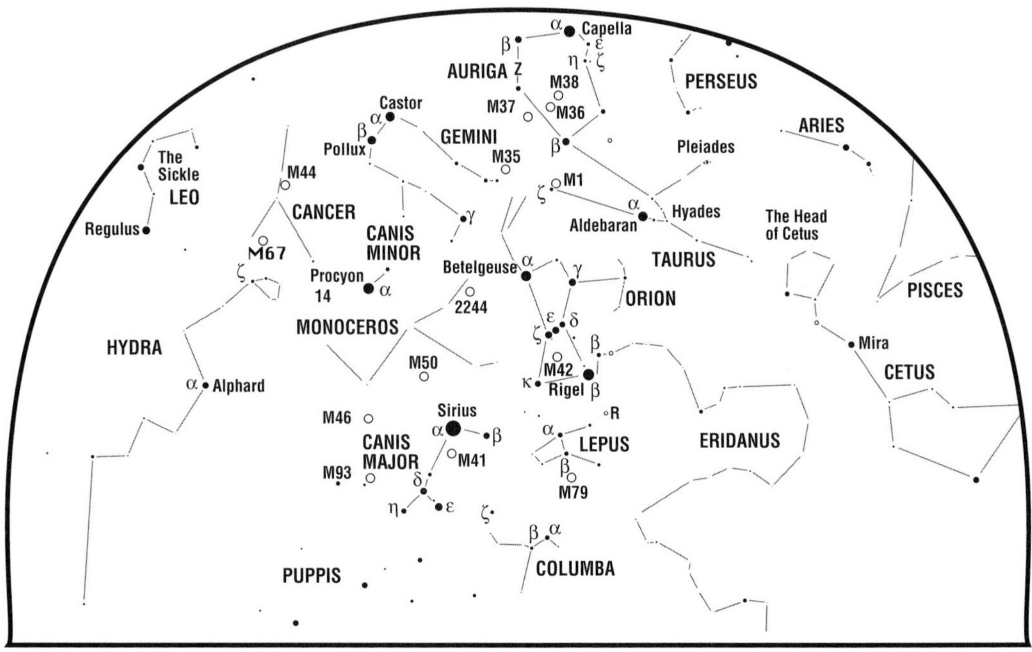

Constellations of the Southern Winter Sky
Canis Major Canis Minor Columba Eridanus Gemini Lepus Monoceros Orion Puppis Taurus

hand, while from his left arm hangs a shield made from a lion's skin. Before him stands Taurus, the Bull, with lowered head and gleaming reddish eye. Actually, it doesn't take much imagination to see this age-old combat in the skies, for the figures are well outlined with stars.

Marking the right shoulder of the giant, Betelgeuse is the only lst-magnitude star that varies in its brightness. This irregular fluctuation amounts to only half a magnitude and probably is a result of the star's extreme age. Betelgeuse is thought to be a dying sun, one of the class of stars known as red giants, an especially fitting title in this case since it is a giant in actual size and its ruddy hue is quite apparent to the naked eye. In 1920 Betelgeuse became the first star whose diameter was directly measured. This was accomplished with the aid of an instrument called an interferometer attached to the 100-inch Hooker telescope on Mt. Wilson, California, then the largest in the world. Subsequent measurements show the star is between 480 and 800 million miles in diameter, or 550 to 920 times the size of the Sun.

Diagonally opposite Betelgeuse in the figure of Orion is Rigel, the constellation's brightest star. It, too, can only be talked of in superlatives, for it is one of the most remote of the visible stars as well as one of the hottest. Its distance is 1,000 light-years, meaning that light, which can encircle

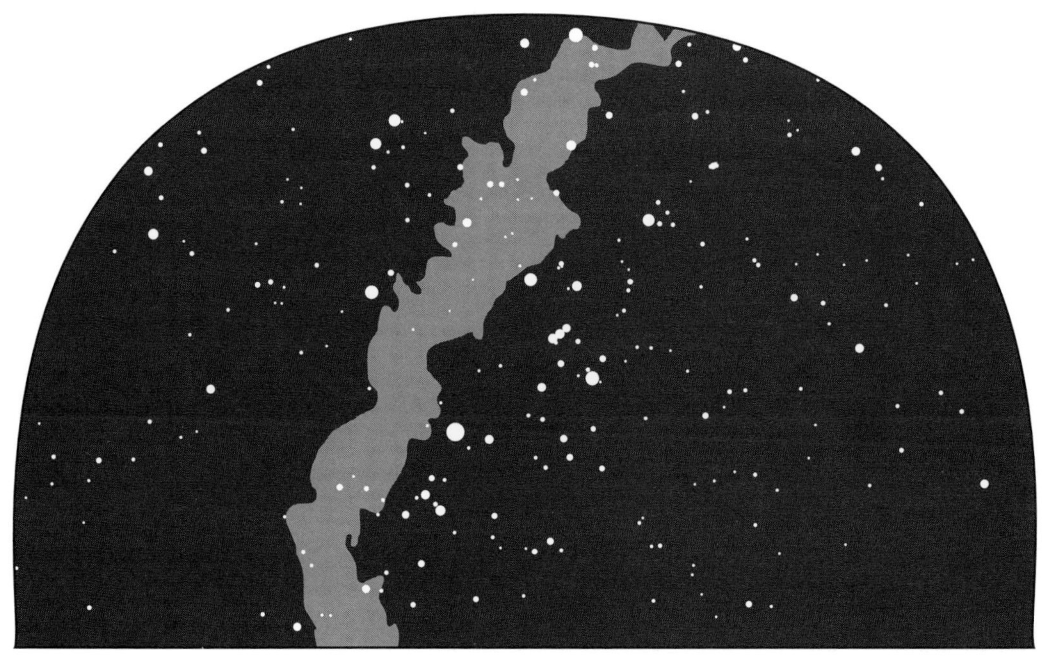

The Southern Winter Sky As Seen On:

December 15 at 12 a.m.	*January 15 at 10 p.m.*	*February 15 at 8 p.m.*
January 1 at 11 p.m.	*February 1 at 9 p.m.*	*March 1 at 7 p.m.*

the Earth seven times in a single second, requires 1,000 years to reach us from this star. Yet even at this distance, Rigel ranks seventh in order of brightness of all our stars. It would take the combined light of 57,000 Suns to equal that of Rigel.

Midway between Rigel and Betelgeuse, three 2nd-magnitude stars form a nearly perfect line just three degrees in length. This is Orion's Belt. So striking and unusual is this three-starred figure that it serves as a means of identification for the constellation. Mintaka, the upper star of the Belt, is exactly on the equator of the sky. Therefore, if you were standing at our North Pole, this star would be precisely on your horizon, where it would mark out the southern boundary of your circling hemisphere of stars. Mintaka is a double star which 10x binoculars will split, but it still conceals its 7th-magnitude companion from my 7x binoculars.

Hanging downward from the Belt is Orion's Sword. Even with your naked eye, you will note a haziness about the central star of this small shape. As the poet Tennyson writes of it in "Merlin and Vivien":

A single misty star,
Which is the second in a line of stars
That seem a sword beneath a belt of three.

I never gazed upon it but I dreamt
Of some vast charm concluded in that star
To make fame nothing.

That "single misty star" is the Orion Nebula, the finest of its type visible from the Northern Hemisphere. It is a vast cloud of glowing gas and dust illuminated from within by a number of embedded hot, blue stars. Binoculars greatly increase the visible size of this cloud in which only its brightest star, Theta (θ), can be seen with low power.

Designated as M42, the Orion Nebula is some 2,800 light-years from us, and light would require 30 years to cross this nebula from side to side. Oddly enough, Galileo, who made so many first sightings with his new telescope, missed the Orion Nebula completely, though it was well within the powers of his tiny optic tube. More than two hundred years after Galileo's time, M42 was watched by stargazer Lord Tennyson, who often "gazed upon it" with his own 2-inch telescope.

Orion is a constellation of so many visual delights that you won't be limited by the dates and hours that I have listed here. They are for the times when he makes his best appearance, high in a dark sky well above the haze of the horizon. As you get to know this sparkling giant, you will come to watch for him as he rises in the east at about 10 o'clock on a frosty night in late October. As the weeks go by, you can follow his slow march across snowy winter landscapes to his final setting in the balmy twilights of late April.

TAURUS

Sighting upward along a line drawn through the three Belt stars of Orion leads the eye close to rosy-red Aldebaran in the constellation Taurus, the Bull. Aldebaran's lst-magnitude glow represents the gleaming eye in the lowered head of the legendary figure. Here, within a small sky area, you'll find three bright star colors. Through your binoculars, look long at rose-colored Aldebaran, then swing to red Betelgeuse, then to blue-white Rigel, then, finally, back again to Aldebaran. It will seem as if you are viewing a display of garnets, rubies, and sparkling diamonds.

Aldebaran glows amid a star figure shaped like a letter V that forms a scattered open cluster called the Hyades. It is known to be a moving cluster, for all of its principal stars, with one exception, are speeding through space as a group. This suggests that, at some remote time and place, they may have started out together. The cluster is 150 light-years from us, moving in the direction of Betelgeuse in Orion. Because of his superior brightness, you might assume that Aldebaran is the leader in the Hyades hegira, but he is not even a member of the clan. Actually, Aldebaran is 70 light-years off, less than halfway to the cluster, which merely furnishes a sparkling background for the Bull's-eye star. Your naked eye can readily make out a couple of wide doubles in the outline of the V, and binoculars add scores of brilliant objects to the glittering field.

Curving upward from the lowered head of Taurus, his horns are tipped with two bright jewels, 3rd-magnitude Zeta and 2nd magnitude Beta (β) or El Nath, which is also used to complete Auriga's pentagon. Near Zeta is M1, the Crab Nebula, the hazy remains of a star that exploded in A.D. 1054. It was number one in a list of 103 diffuse objects catalogued by Charles Messier, the great comet hunter, in 1781. You may just glimpse it with a low power telescope as a fuzzy starlike speck. It is properly a telescopic nebula, but there are many other Messier objects (such as Orion's M42) that are within the range of your binoculars, and we will locate and discuss them as they come into view.

I have saved the best of Taurus until last, and now it seems almost a waste of time and effort to tell you where to locate it in the sky, for I have yet to find the person who has never seen the Pleiades (M45) or who didn't express amazement and delight at their appearance in binoculars. It is the finest open star cluster in the sky.

This small group is often known as the Seven Sisters, and six of them are quite easily seen with the naked eye, even with some moonlight present. On the clearest of nights, I have often seen nine stars in this grouping, though some earlier observers have reported as many as fourteen naked-eye stars in the cluster. Binoculars, of course, reveal dozens more, though I can find no colored stars among them. This isn't remarkable, as the Pleiades—like the Hyades—form a moving cluster, so a common star type should be expected.

The Pleiades lie only some five degrees north of the ecliptic, and therefore the Moon, in its travels around the Earth, occasionally moves between us and the cluster causing some of its stars to be occulted, or blocked. When conditions are favorable, several of these occultations can occur in the course of a few months. If you are fortunate enough to witness one of these performances you may be surprised to find that the Moon, which can seem so large when near the horizon, fits quite easily into the bowl-shaped figure of the little cluster. This entire group of stars is festooned with wisps of the nebulous matter from which its stars were formed. Its wraith-like aspect shows up best in long-exposure photographs, though you can sometimes get vague glimpses of it with low-power binoculars.

ERIDANUS

Close beside bright Rigel in Orion lies the source of Eridanus, the River, a name doubtless suggested by the meander of a stream of rather faint naked-eye stars that seems to first flow westward, then turns and drops to the horizon to finally end in the gleam of lst-magnitude Achernar, a star too far south for us to see in mid-northern latitudes. There is little about this river to encourage any fishing with binoculars.

LEPUS

Just beneath Orion crouches Lepus, the Hare. The ears of this timid creature appear as two wide pairs of stars below Rigel. Gamma (γ) is an easy double, magnitudes 3.7 and 6.3, that is 96" apart. The globular cluster M79 appears as a bit of haze in your binoculars. To the west, in the location marked R, is the famous variable star R Leporis. This deep crimson star will be described and charted in a later chapter on variables.

COLUMBA

Midway between Lepus and the south horizon is Columba, Noah's Dove. There is nothing of interest for our binoculars in this small four-starred figure, but I will mention it here to give you one more species to add to your Christmas bird count.

CANIS MAJOR

If you follow a straight line from Aldebaran to Orion's Belt onward for a similar distance, it will point out Sirius, the Dog Star, in the constellation Canis Major, the Greater Dog. Sirius is by far the brightest star in the sky with a magnitude of -1.5, which makes it twelve times brighter than Aldebaran. It is also nearer to Earth than any of the other naked-eye stars visible from our latitude, and this is largely responsible for its great brightness. Light requires only 8 years and 9 months to reach us from Sirius, and this—compared to the 1,000-light-year distance of Rigel—puts the Dog Star almost on our doorstep.

Sirius is a binary star with a most remarkable companion star called Sirius B that circles about it in a period of 50 years, but without a telescope of fair size, it is usually drowned in the bright glow of the primary body. Scarcely larger than three times the diameter of Earth, this tiny star is composed of material so dense that a teaspoonful of it would weigh a ton. If you weigh 150 pounds here on Earth, you would weigh 2,000 tons on the Pup, as Sirius B is often called. This miniature star is a white dwarf. It probably is the remaining central portion of a once-large star that has used up all its hydrogen and collapsed around its heavy nucleus.

To me, Sirius seems to flicker more than any other star. It is a white star, but because it is always rather low in our sky, thick layers of atmosphere cause it to throw off flashes of vivid color that are beautiful to watch in binoculars.

Canis Major contains four 2nd-magnitude stars which by themselves would make this a sparkling constellation were they not so overwhelmed by brilliant Sirius. The ancients believed that when the Greater Dog rose with the Sun in late summer, the heat of its stars, added to that of the Sun, brought on the sultry period known as "dog days."

Look at the open cluster M41 a few degrees below Sirius. It can be seen with the naked eye on clear, moonless nights; your binoculars will expand it into a lovely company of glittering stars.

PUPPIS

You may notice a few 2nd- and 3rd-magnitude stars lying between Canis Major and the southern horizon. They belong to the constellation Puppis, the stern of the ship Argo of Jason's legendary Argonautic Expedition. Most of the remainder of the ship lies below the horizon and can't ever be seen from our latitudes—which is unfortunate, as this excludes our seeing Canopus, the second brightest of all the stars.

The Milky Way flows through Puppis, making fine sweeping for binoculars, with many rich fields and open clusters in the northern part of the region. Of these, M46 and M93 are the most rewarding, the latter having a diameter of about half a degree and containing numerous resolvable stars. The star Pi (π) lies in a fine field with bright companions.

CANIS MINOR

The first bright star east of Betelgeuse in Orion is Procyon, in the constellation Canis Minor, the Little Dog. Appropriately, it is a small constellation. Just above Procyon is the 3rd-magnitude star Beta with its two small attendant stars making a tiny triangle. Like Sirius, Procyon is a neighbor of our Sun, only 11 light-years distant. Oddly enough, it also has a faint white dwarf companion star circling about it, this one with a period of 40 years.

The star 14 Canis Minoris is an optical triple star, which you can catch in binoculars when conditions are just right. Its magnitudes are 5.4, 8.4, and 9.3. See the key chart for the location of 14 CMi.

MONOCEROS

Within the triangle marked by Betelgeuse, Sirius, and Procyon lies most of the constellation Monoceros, the Unicorn. Even though the winter Milky Way runs directly through here, there are no bright stars to attract the eye, and the outline is difficult to trace. There is, however, some rich sweeping here for binoculars, and some interesting clusters may be found. Of the latter, look at NGC 2244 clustered around a 6th-magnitude star and the open cluster M50 about midway between Alpha and Beta.

GEMINI

To the north of Procyon you will find the two bright stars, Castor and Pollux, in the constellation Gemini, the Twins. There is much of interest here. In mythology, Castor and Pollux were twin brothers who accompanied Jason on the Argonautic Expedition. Although these two stars appear in the sky to be somewhat closer together than the Pointers in the Dipper, there is no real relationship between them, for Castor is 13 light-years more distant from us than Pollux. Castor, in fact, has no need for any family ties with Pollux; he is the brightest member of a family of his own—a system of six stars revolving about their common center of gravity. Many years ago, I could readily divide Castor and his 3rd-magnitude companion with my two-inch spyglass, but they have gradually drawn closer and this is no longer possible. There is a noticeable difference in brightness between these twins: Pollux is a faint 1st magnitude star, while Castor is a bright 2nd magnitude.

Eta (η) Geminorum is classified as a red giant, an old star similar to Betelgeuse in Orion. It also varies from 3rd to 4th magnitude in a period of about eight months. The star Zeta is a fine double in binoculars and is a variable as well, with a range of half a magnitude in about 10 days. At its brightest it compares with Delta (δ) and Lambda (λ).

The star cluster M35 is bright enough to be seen with the naked eye on clear, moonless nights, though binoculars will add greatly to its interest and beauty. Just west of M35, a 5th-magnitude star marks the spot occupied by the Sun when it is farthest north at the beginning of summer (June 21). It was very close to this same spot in Gemini that, on the night of March 13, 1781, the English musician and amateur astronomer William Herschel, using a homemade reflecting telescope, came upon an object that showed a tiny disk rather than a starlike point. At first he thought it was a comet, but later observations showed the broad elliptical orbit typical of a planet. It eventually was named Uranus—the first new member to be added to the family of the Sun.

Gemini was once again to figure in astronomical history early in 1930, when American astronomer Clyde Tombaugh found the tiny planet Pluto near the 4th-magnitude star Delta Geminorum.

This ends our survey of the winter stars. Those constellations you may have noted far to the west on the winter sky maps will be treated later as autumn stars, after they have risen in the east and are well positioned for observing. The stars along the eastern border of the southward-looking winter map will become the stars of spring and, together with the northern skies of the vernal season, they now claim our full attention.

6
The Spring Stars

FACING NORTH

The sky maps and key charts in the following pages show the skies as they appear three months later than the maps and charts of the winter season shown in Chapter 5. Since one-fourth of a year has passed since then, the stars pictured here have made one-fourth of a revolution about the polar point. The Big Dipper, which in winter stood on its handle directly east of Polaris, is now almost overhead and at its highest point in the sky. Of course, all the other stars have also pivoted a quarter turn around the pole.

Now, low in the eastern sky, stars that will light our summer nights are rising, while those we viewed high in our winter skies are now setting in the west. The Milky Way that was so bright before now lies all along the northern horizon and often is completely blotted out by haze. Only one star of all those pictured seems to have held its place—Polaris.

But even it has not always stood so close beside the pole as it does today—nor will it always remain so in the future. All things change with time, and pole stars are no exception. If you could have stood somewhere upon a hilltop and watched, as in a time-lapse sequence, the age-long pageant of the slowly changing skies, you could have witnessed, as the millennia marched by, an awesome spectacle that no earthly eye has ever seen—the migration of the stars.

Just as most of our familiar birds fly north or south according to the seasons of the year, so do the stars, as seen from Earth, move north and south when prompted by the passing of another and much longer year—a period 25,800 Earth-years long, sometimes referred to as the Great Year.

Both of these mass movements, of birds and of stars, are in obedience to the dictates of the polar axis of the Earth. It is the 23½-degree tilt of this axis that gives the Earth its seasons and thus provides the birds a motive

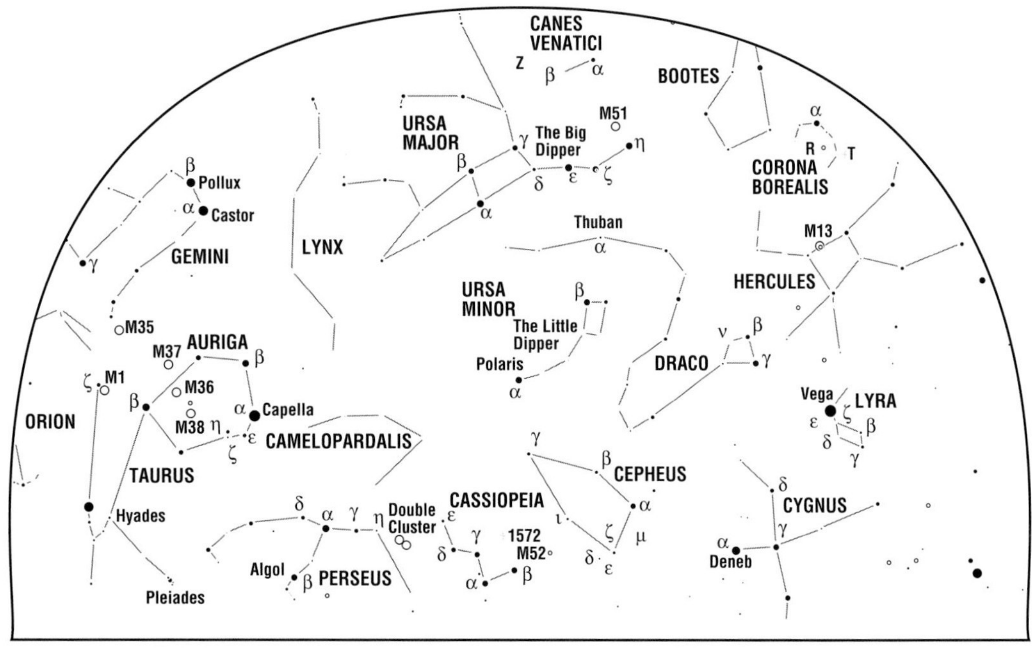

Constellations of the Northern Spring Sky
Camelopardalis Cepheus Draco Gemini Hercules Ursa Major

for migration. And another feature of this axis, a slow wobbling motion or gyration, causes the apparent migrations of the stars.

The Earth is not a perfect sphere. It is like a ball that is slightly flattened at its poles and bulges out at its equator. This distortion is sufficient to make its equatorial diameter 27 miles greater than its polar diameter. This equatorial bulge is tilted 23½ degrees to the ecliptic plane in which the Earth revolves about the Sun. In obedience to the law of gravity, both the Sun and Moon are unceasing in their efforts to pull this encircling bulge into their common plane. Fortunately for bird watchers and stargazers and all who appreciate the changes the varied seasons bring, the rapidly spinning Earth acts as a gyroscope and is able to resist this pull—except for such side effects as the wobble of its axis.

At present, our polar axis points northward toward the star Polaris. It has been our pole star for hundreds of years, though its reign is only temporary. The slow gyration of the axis causes it to point to a succession of different pole stars. All lie near a circle 47 degrees in diameter that the wandering pole traces out upon the sky. Any fairly bright star that lies near this circle serves as the pole star for a portion of the Great Year. The drawing on page 36 shows this circling path among the past and future pole stars of the northern sky.

The Northern Spring Sky As Seen On:
March 15 at 12 a.m. *April 15 at 10 p.m.* *May 15 at 8 p.m.*
April 1 at 11 p.m. *May 1 at 9 p.m.*

I have often thought how fortunate we are to have such a conspicuous pole star as Polaris, one we can find so easily by just sighting along the Big Dipper's Pointer Stars. Five thousand years ago, when the Pyramids were being built, the pole star was Thuban in the constellation Draco. It is much fainter than Polaris and is located about midway between Mizar in the Big Dipper and Kochab in the Little Dipper's bowl.

Thuban serves well to illustrate the north and south migrations of the stars. In that long season of the Great Year when Thuban reigned as the pole star of the sky, the Southern Cross rose high enough above the southern horizon to be seen from as far north as Quebec, in Canada. Today it is so far south that, in this country, the entire Cross can be seen only from the southern tips of Florida and Texas.

The true pole now is a little more than two Moon-diameters distant from Polaris and is slowly drawing closer to it. In the year 2095, it will be at its nearest—less than one Moon-diameter away. The pole will then slowly move away in a counterclockwise direction. For ages it will creep among the stars of Cepheus, first passing 3rd-magnitude Gamma Cephei around the year 4000 at about half the distance between the Pointer Stars. Then, a couple of millennia later, Alpha Cephei, or Alderamin, will become another distant pole star.

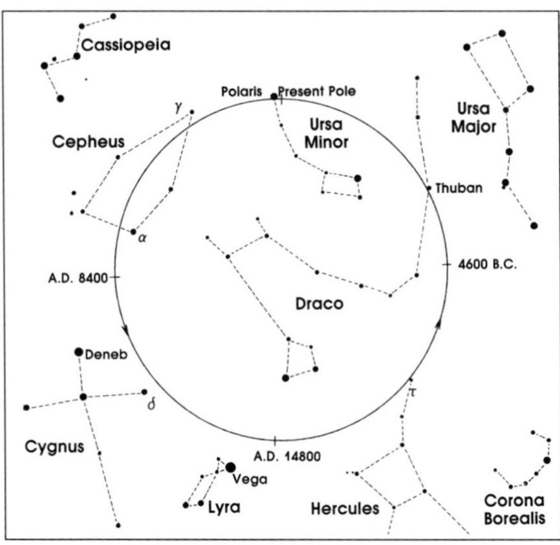

Pole Stars of the Great Year

Some 12,000 years from now, the polar point will have completed almost half its circle in the sky, and blue-white Vega will be the brightest of all pole stars, even though it will be some five degrees distant from the true pole. The skies themselves will not have changed much in that time, save in their relation to the pole. We still would recognize all our familiar stars and constellations. The Big Dipper still will be a dipper, and Castor and Pollux, the Twins of Gemini, will still move side by side across the sky.

However, with the polar point near Vega, the habits and the seasonal associations of the stars and constellations will have vastly altered. No longer will Polaris and the Big Dipper be circumpolar, but instead they will rise and set. No longer will the Twin Stars of Gemini ride high above our heads on winter nights. They will then be features of late summer evenings and will cross the sky just above the southern horizon. Missing completely from those distant future skies of our mid-latitudes will be Sirius, the brightest star of all. The dazzling Dog Star, along with blue-white Rigel and the Belt Stars of Orion, will then be on their southern migration.

The periodic migration of the stars is a vast cycle far more certain and predictable than the coming of the swallows to Capistrano or the mid-March meeting of turkey buzzards at Hinckley Ridge, Ohio. Though no human eye can note its flow, the passing of the Great Year with its migrating stars is a repeating process. At some distant future date, the polar point will creep past Thuban and then, ages later, the star Polaris will slow its spiral circling to once again become the Pole Star.

FACING SOUTH

Springtime is the most exciting season of the bird-watcher's year. The three months March, April, and May bring the birds back from the south again. Through many of these early days, the winter-weary eye is searching for the first sight of a robin hopping on the lawn, while all the time the ear is bent to catch the first trial birdsong tune-up notes that, as the days go by, will be augmented to a full-fledged chorus.

The stars are even more responsive than the birds to the never-ending succession of the seasons. The rising of the stars in the east and their setting in the west is so punctual that the world's chronometers are set according to their crossing of the sky. Already you can see this relentless tide in action. Go out just after dark on a mid-April night and look well down in the western sky. There you will see, sinking lower night by night, the entire coterie of bright stars that you watched on midwinter nights when they were high in the southern sky. As you view this farewell party on succeeding nights, you will note that Rigel in Orion is the first to go, followed by the Pleiades and then Aldebaran in the Hyades. Next to leave is Sirius which, as it nears the horizon, often shows a final flash of green or red. Then Betelgeuse, Procyon, Capella, and Pollux disappear in order as they drop behind the Earth. But the departing winter stars have been replaced by a lively company of spring stars and constellations. They now invite our close inspection, for often they will tax the powers of our probing binoculars.

LEO

Looking just to the west of the meridian and well up in the sky on a midseason night, you can't miss 1st-magnitude Regulus in the constellation Leo, the Lion. According to the older star charts, Regulus marked the heart of the Lion, but for us it gleams at the end of the handle of the Sickle. As with the Big Dipper, this is a realistic figure.

With your binoculars, note the difference in color between white Regulus and deep yellow Gamma Leonis in the end of the Sickle. Within the sweeping curve of this Sickle lies the radiant point of the Leonid meteor shower that occurs each November. There will be more on meteors later in a special chapter. The star Zeta, just above Gamma, will show three companions nearby if your eye is keen and your binoculars steady. Epsilon (ε), which is also in the Sickle, makes a little triangle with two 7th-magnitude stars. There is no physical connection involved here.

The letter R on the chart marks the location of a well-known variable star. At regular intervals this star becomes bright enough to be seen with 7x binoculars, and I have often seen it with the naked eye. So important are variable stars and as easily observed are many of the brighter ones that I have devoted a later chapter to the systematic study of them. Numerous charts have been included in that chapter showing just where to find them in the sky with your low-power binoculars.

Directly east of the Sickle, you will find a conspicuous triangle of stars, the brightest of which is Beta (β). This star is Denebola, a name meaning "The Tail of the Lion." Below this triangle, the star Tau (τ) is a fairly easy double in binoculars. Denebola and Gamma, which is in the Sickle, are

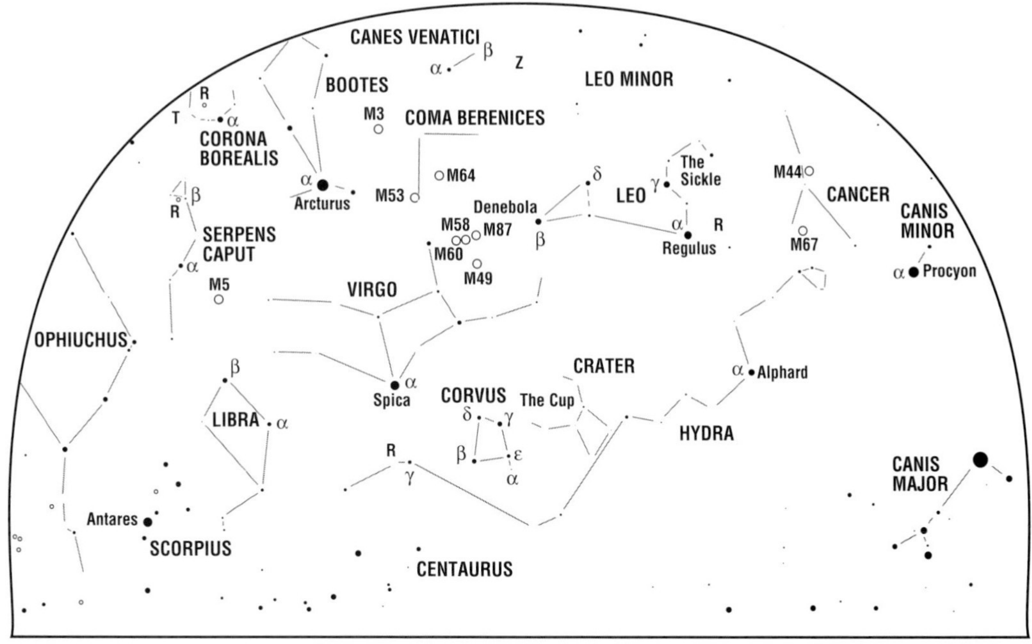

Constellations of the Southern Spring Sky

Bootes   Cancer   Canes Venatici   Coma Berenices   Corona Borealis
Crater and Corvus   Hydra   Leo   Virgo

both standard 2nd-magnitude stars, with Gamma only 0.15 magnitude brighter. It takes practice to detect magnitude differences this small, as you will see when observing variable stars.

Regulus lies almost directly on the ecliptic, the center line of the zodiac, so sometimes Leo is a busy place. All the planets pass through here on their trips about the Sun. If you find a bright stranger in this region that is not shown on a chart, it will doubtlessly be one of the brighter planets. The Moon also travels through Leo in its swing around the Earth, and it will sometimes occult, or hide, some of Leo's stars from your view.

CANCER

Regulus and the Sickle lie in a rather barren sector of the sky, and this makes them stand out in greater contrast so they serve as beacons to guide us to a number of fainter outlying stars and constellations. For example, halfway along a line from Regulus to Pollux is the tiny constellation Cancer, the Crab. On moonlit nights, you may have trouble finding it at all with your naked eye, though your binoculars make it easy. At the spot marked M44, you can find the open star cluster known as Praesepe, or the Beehive, with your naked eye on a sparkling, moonless night. Your binoculars will show that this is not just a hazy cloud but a swarm of tiny

The Southern Spring Sky As Seen On:
March 15 at 12 a.m. *April 15 at 10 p.m.* *May 15 at 8 p.m.*
April 1 at 11 p.m. *May 1 at 9 p.m.*

stars. Galileo, with his crude new telescope, was the first to learn this. He counted thirty-six stars in the group. Closely preceding the star Alpha is the dim star cluster M67. It is difficult in 7x binoculars, for it only shows as a tiny bit of haze.

HYDRA

North of Cancer and the Sickle are the constellations Lynx, the Lynx, and Leo Minor, the Lesser Lion. These are small, faint figures with nothing of importance to offer us. Directly south of Cancer, or about halfway between Regulus and Procyon, look for a compact group of five 3rd- and 4th-magnitude stars. This conspicuous figure forms the head of Hydra, the Water Snake. In total area, it is the largest of all the constellations, for it stretches its sinuous folds across the southern springtime sky with its tail now low in the southeast. Hydra's brightest star is 2nd-magnitude Alphard, known as the Solitary One because there is no other bright star nearby. Alphard, like Betelgeuse in Orion, is classed as a red giant. Use your binoculars to bring out the redness of its fading fires. Another red star, the variable R Hydrae, is charted and described in a later chapter.

Just east of Alphard and beneath Regulus is a small figure or asterism known as Sextans, the Sextant. It has no star brighter than 5th magnitude.

Sextans is a modern invention designed by Hevelius in 1690 to fill in sky space not occupied by the adjacent constellations.

CRATER AND CORVUS

Resting on the coils of Hydra directly in the south at this time are two small figures of ancient lineage. The westernmost of these is Crater, the Cup, a pretty design made up of faint stars in the outline of a cup or goblet. Binoculars will help bring out its stars. To the east of Crater, Corvus, the Crow, is brighter and more interesting, though there is little about this celestial crow to remind you of the *Corvus brachyrhynchos* of your bird guide. My old atlases show this raven facing westward and pecking at the coils of Hydra on which it stands. The stars Delta and Gamma Corvi are in the shoulders and Alpha is in the beak. Delta is a fairly easy double.

VIRGO

Like the Big Dipper, Corvus has its pointer stars. The two stars in its shoulders point to a nearby bright star, lst-magnitude Spica in the constellation Virgo, the Virgin. Like Leo, this large constellation lies in the zodiac, and thus the Moon and planets all pass through here. So, too, does the Sun, for just west of the star Eta is the location where the Sun stands when the days and nights are equal at the beginning of the autumn season.

Within the rough square formed by Epsilon, Gamma, and Beta in Virgo and Denebola in Leo lies the so-called Realm of the Galaxies. Here a telescope reveals a profusion of galaxies that, because of their great distance from us, are mostly too faint for our low-power binoculars, although you can catch glimpses of a few, such as M49, M58, M60, and M87.

COMA BERENICES

Just to the north of the Realm of the Galaxies, which we saw fenced in by Denebola and three of Virgo's stars, is the faint constellation Coma Berenices, or Berenice's Hair. Look for this widely scattered cluster of faint stars on a clear, moonless night. You will make out perhaps a dozen individual stars with your naked eye, while binoculars—preferably low power and wide field—will add many more. This whole region is actually a continuation of the Realm, for photographs of this area show that it, like Virgo, is strewn with thousands of galaxies, all of which are millions of light-years distant.

With my 6-inch telescope I have often seen several of these faint galaxies in a single field of view. When comet-hunting, I avoid this Coma-Virgo-Leo region because of the abundance of these objects. So many of them look just like faint comets that to check them out for possible movement would take too much time.

In legendary lore, Berenice was the queen of ancient Egypt. When the king left to battle the Assyrians, she vowed to dedicate her hair to the goddess of beauty in the hope that this act of sacrifice might assure his safe return. Accordingly, her tresses were duly shorn and deposited in the Temple of Venus, but soon they disappeared and could not be found. The King came home from battle victorious and unharmed but, understandably, became quite angry to see that Berenice's hair was gone—not only from her once queenly head, but also from the temple. Fortunately, the court astronomer was both a stargazer and a diplomat. To console the distraught royal couple for their loss, he took them outside on a clear, dark night and pointed out the dimly lit area of sky where the missing locks had been placed among the stars by Jupiter.

With your binoculars you still can see that hair today, in scattered curls and ringlets, all across the constellation that was named in honor of the queen. To some, of course, those stolen locks look more like curving lines of twinkling stars.

Near 4th-magnitude Alpha, your binoculars should easily locate the globular cluster M53, while about four degrees to the northwest of Alpha near a 5th-magnitude star you will find the small galaxy, M64, called the Blackeye Galaxy.

CANES VENATICI

Directly north of Coma Berenices is a tiny, modern star figure known as Canes Venatici, the Hunting Dogs. You will find this dim constellation as three stars creating an obtuse triangle just beneath the bend in the handle of the Big Dipper. The brightest of these is 3rd-magnitude Alpha or Cor Caroli. About three degrees south of the end star in the Dipper's handle, some patient searching may reveal the galaxy M51—the famous Whirlpool Galaxy, the first galaxy observed to be spiral in form. In binoculars it will only show as the faintest bit of haze. The globular cluster M3, beside the southern border of the constellation, is much brighter, however. I can always locate it with the naked eye. Low-power binoculars give it a bright nebulous appearance but will not resolve its stars into separate points.

BOOTES

A line drawn from Beta through Alpha in Canes Venatici leads the eye almost directly to Arcturus, the brightest star north of the sky's equator. Of the fifteen 1st-magnitude stars visible in our latitude, only Sirius, the winter star, is brighter. Another way to find Arcturus is simply to follow the curving arc of the Dipper's handle to the first bright star, which is Arcturus. Continue this same curve for an equal distance and you next arrive at Spica in Virgo, which Corvus, the Crow, has already pointed out.

Arcturus stands at the southern tip of a kite-shaped figure called Bootes, the Herdsman. This brilliant yellow star—more than a hundred times brighter than our Sun—is classed as a red giant, though it is not as far advanced in its evolution as is ancient Betelgeuse. Arcturus moves through space more rapidly than any of the bright stars we see. This speed is estimated at about 200 miles per second across our line of sight.

You may feel you should grab those binoculars quickly and dash outside to get a look at old Arcturus before he leaves the sky completely. But you can take your time on this, for Arcturus has been bowling along at that rate for millions of years. In the last two thousand years, it has only moved across our line of sight about one degree, or twice the apparent diameter of a Full Moon.

Arcturus is 36 light-years distant from the Earth. This was dramatized in 1932 at the opening of the Century of Progress Exposition in Chicago by utilizing the light from the star to activate a switch that turned on all the exposition lights. Arcturus was given this stellar role because at the time astronomers felt the star was 40 light-years away and that its light had started toward the Earth 40 years before—at the time of the Columbian Exposition in 1892.

CORONA BOREALIS

Closely following Bootes, the Herdsman, in his march across the skies, you will find the small but beautiful constellation Corona Borealis, the Northern Crown. The 2nd-magnitude star Alphecca (α), the so-called Pearl of the Crown, is the brightest star of the entire figure. Within the curve of the horseshoe-shaped outline of the Crown is the important variable star marked R.

Just outside the circlet of stars, a cross marks the location of T, a new star that rose to 2nd magnitude in 1866 and then gradually faded back to 9th magnitude. Eighty years later, in 1946, it suddenly rose to 3rd magnitude and is now somewhat restless at 10th magnitude. Here in this tiny constellation are two important stars, the variable R and the recurrent nova T, only about four degrees apart. You should check these two locations each clear night; to help you, I have included charts and fuller details in later chapters.

7
The Summer Stars

FACING NORTH

Gone completely from our balmy nights is the glitter of the winter stars. Gone are the giant striding figures of Orion and the charging Bull. The bright Sickle of our springtime skies is also missing, and speeding Arcturus now is slipping toward the west. Gone, too, for the season are the frigid nights, the frosted lenses, the heavy clothing, and the mittened hands. In place of winter's rigors and the sometimes gusty nights of spring, the summer skies now offer you and your binoculars complete observing comfort and relaxation as you survey, lying on the ground or steadied by a chair, a horizon-to-horizon floodtide of celestial sights.

DRACO

In the north, high above Polaris and almost encircling the Little Dipper, the ancient constellation Draco, the Dragon, holds its four-sided head almost directly above you. Some say that this figure really does look like its namesake, but since I've never seen a dragon I find instead a giant measuring-worm, its back looped high above it and its head reared back as it contentedly munches on some juicy greenery from my garden. The four stars of its head are of four different magnitudes: 2.5, 3, 4, and 5; therefore they, like the Little Dipper stars, furnish us a scale of comparison for estimating other stars. Look very closely at Nu (ν), the faint star of the head. Keen eyesight can split this star into two components, both 5th magnitude and just 1' apart. This is really a binary system that is easy to find with binoculars. So distant are these two from each other that at least a million years is required for a single common revolution.

Earlier we paid passing tribute to the third star in the Dragon's tail. This is Thuban, the alpha of the constellation, though it is by no means its brightest star. It can be found about halfway between the bowl of the Little

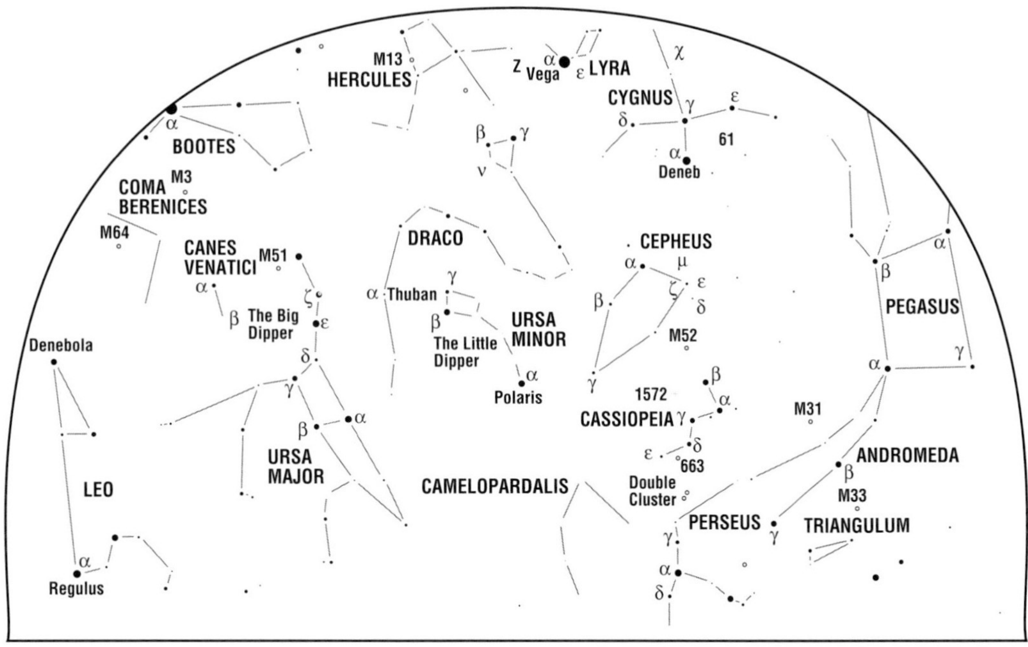

Constellations of the Northern Summer Sky
Camelopardalis Cassiopeia Cepheus Draco Ursa Minor

Dipper and Mizar in the handle of the Big Dipper. Thuban was the Pole Star of the sky at the time of the building of the Pyramids some 5,000 years ago. One wonders if the Pyramids will still be standing when Earth's slowly shifting axis swings back to Thuban once again in A.D. 21,000.

CEPHEUS

Just east of Polaris is an inconspicuous five-sided figure made up of 3rd- and 4th-magnitude stars resembling the end view of a house with a steep-pitched roof. This is Cepheus, the King. For all its faintness, Cepheus has some interesting features and, as the Milky Way passes here, the entire region deserves sweeping with your binoculars. Just below the base of the "house," note the faint star Mu (μ). It is often called the Garnet Star, for it is the reddest naked-eye star to be seen from this latitude. In order to fully appreciate its color, get Mu and nearby Alpha, which is a white star, in the same field of view in your binoculars. The contrast of colors is striking. Mu is also a variable star, somewhat like Betelgeuse in Orion, and fluctuates between magnitudes 3.4 and 5.1 over an irregular period.

Delta Cephei is both a double star and a variable star. Its companion glows at magnitude 7.5 and is blue; Delta is yellowish and varies from 3.5 to 4.4. You can watch this variation from night to night and make your

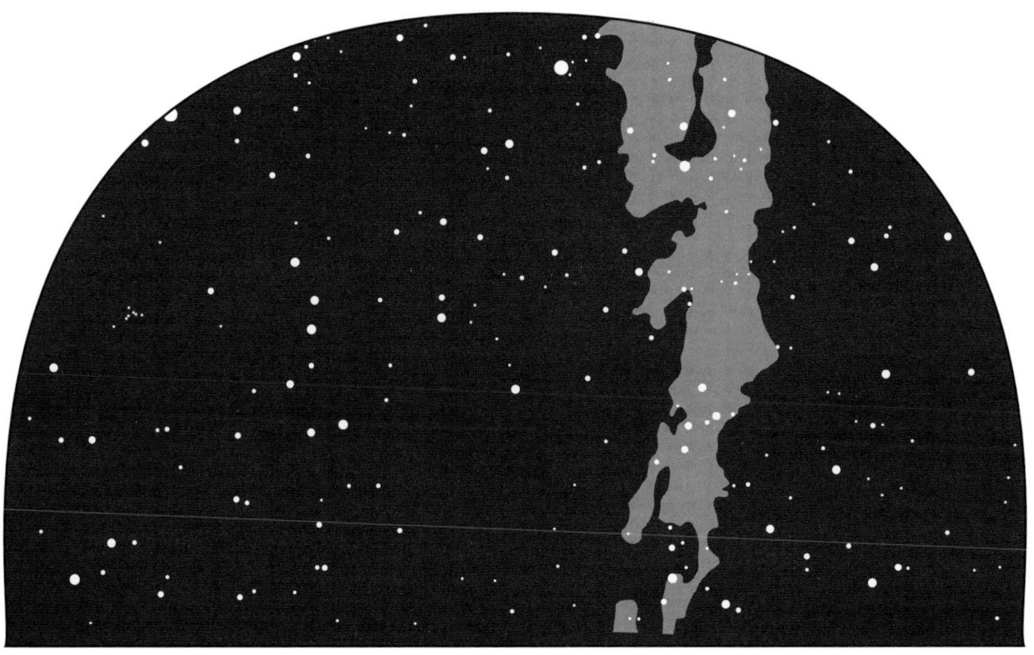

The Northern Summer Sky As Seen On:

June 15 at 12 a.m.	*July 15 at 10 p.m.*	*August 15 at 8 p.m.*
July 1 at 11 p.m.	*August 1 at 9 p.m.*	*September 1 at 7 p.m.*

own determination of its period. You'll find a chart and a more detailed account of this important star in a later chapter on variable stars. This chart also covers the variable Mu Cephei.

CAMELOPARDALIS

The large, virtually empty area extending from Polaris down to the northern horizon is occupied by the constellation Camelopardalis, the Giraffe. This must be one of the dullest of all the constellations, for it has nothing brighter than 4th magnitude. At this time of year you will have trouble even finding these stars—which may be just as well, for there is nothing here of interest for us.

This figure was introduced in 1690 by Hevelius, who also gave us such assorted creatures of the night as the Hunting Dogs, the Lynx, the Lizard, the Little Lion, and a number of others that we will discuss later. Wide open spaces must have appealed to old Hevelius, for his crowning achievement was building a 150-foot-long telescope that had an open tube.

FACING SOUTH

Arcturus, one of the bright stars you watched in spring, is still high in the western sky as you face south to begin your exploration of this

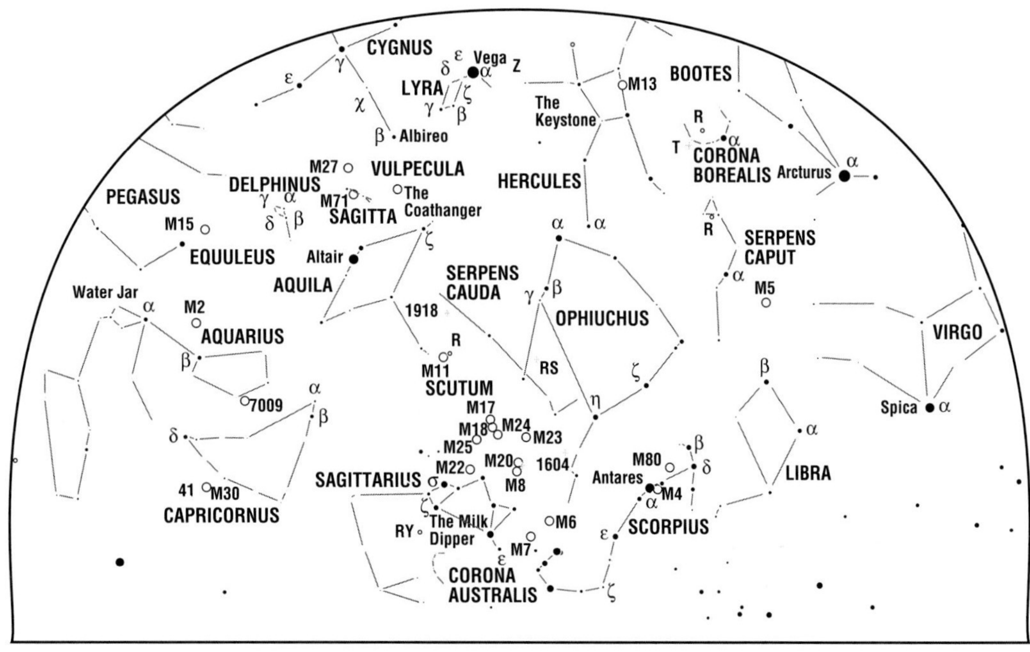

Constellations of the Southern Summer Sky
Aquila Capricornus Cygnus Delphinus Hercules Libra Lyra Sagitta Sagittarius Scutum Serpens and Ophiuchus Scorpius Vulpecula

inviting sector of the sky. Right beside the kite shape of Bootes you will recognize the little circlet of the Northern Crown.

With your binoculars, see if the variable R Coronae Borealis is at its normal maximum brightness of magnitude 5.7. This will only take a couple of seconds and should be a nightly ritual. Sooner or later, this strange star is certain to take a temporary leave of absence from its place within the Crown. In this same field of view also look for the recurrent nova T Coronae Borealis. Although it is almost constant at 10th magnitude and thus a bit beyond the range of 7x binoculars, you never know when it will rise again.

SERPENS AND OPHIUCHUS

This is a mixed-up pair of constellations. Here is the giant figure of a man trying to hold a writhing serpent. As it appears in my old atlas, Ophiuchus occupies the center of the stage with the Serpent's head (Serpens Caput) to the west and its tail (Serpens Cauda) to the east. Ophiuchus should have been in the zodiac, because the Sun, Moon, and planets all spend much more time here than in Scorpius, just beneath it.

Note the five-starred figure that forms the head of Serpens. With your binoculars locate the field of the variable R Serpentis. About once each

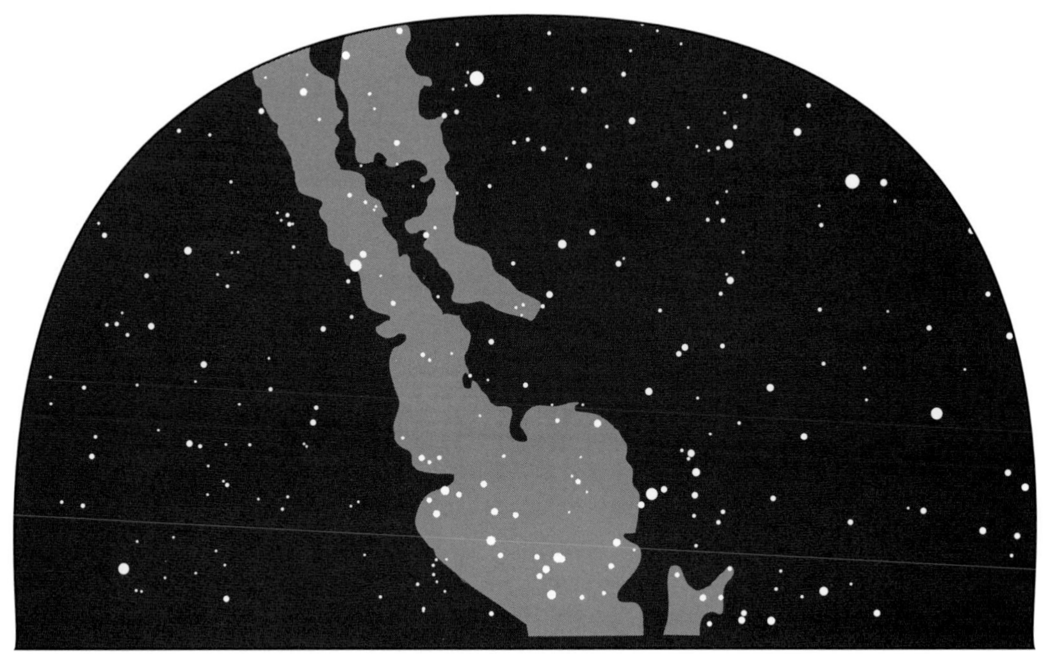

The Southern Summer Sky As Seen On:

June 15 at 12 a.m.	*July 15 at 10 p.m.*	*August 15 at 8 p.m.*
July 1 at 11 p.m.	*August 1 at 9 p.m.*	*September 1 at 7 p.m.*

year it rises to 5th magnitude and can be watched for several weeks before it sinks too low for 7x binoculars. Beta, the brightest star in the Serpent's Head, lies near a brilliant field of stars. The globular cluster M5 can be spotted with the naked eye about five degrees (the distance between the Pointer Stars) southwest of Alpha and right beside a 5th-magnitude star. Near the end of the Serpent's tail, the star Theta is a fine pair of 4th-magnitude stars. About four degrees west of it you will find a beautiful open cluster, IC 4756.

In the past Ophiuchus has produced many novae. At the spot marked 1604 on the sky chart, Kepler's Star blazed out in that year and quickly became brighter than the planet Jupiter. A few months later it had faded from naked-eye visibility. RS marks the location of a recurrent, or repeating, nova. In 1898 it suddenly rose to 4th magnitude and has done the same in 1933, 1958, and 1967. It is presently a restless 11th-magnitude star that could rise again at any time. On the key chart, the location of RS is included so you can watch for another outburst. In 1933 this star brightened about one magnitude per hour.

The Milky Way is a spectacular sight in southern Ophiuchus, and sweeping this region with binoculars will show you many bright patches or star clouds.

HERCULES

High in the sky and not far from your summer zenith is the constellation Hercules, the Kneeler. This is a difficult group, but if you have already found Draco, the circlet of Corona, the Head of the Serpent, and Ophiuchus, you have it more than half-surrounded and should have no trouble with it. Note the four-sided figure that I have marked The Keystone. The little circle near the western edge of it marks M13, the Hercules Cluster. It is faintly visible to the naked eye, and your binoculars will greatly improve the image. M13's million stars still merge into a hazy ball, for they are 23,000 light-years from Earth. Just north of the Keystone your binoculars will pick up a somewhat fainter and more condensed cluster, M92. The star Kappa (κ) is a double that may be within your powers. It can be found a little east of the Serpent's Head.

Alpha is a red giant star—one of the largest known. It is also a variable star that fluctuates in an irregular manner between 3rd and 4th magnitude in a period of about three months.

SCORPIUS

Beneath the feet of Ophiuchus and not far above your southern horizon you can find the bright red star Antares near midstream in the Milky Way. This star marks the heart of Scorpius, the Scorpion, and for a change, this is a realistic figure. It actually does look like a scorpion with its long curving tail trailing along behind it. Like Alpha Herculis, Antares is a red giant. This modern classification is an apt one for it is the reddest of the lst-magnitude stars and its diameter is far greater than the orbit of the Earth. The ecliptic, the center line of the zodiac, passes just about four degrees north of Antares, so the Moon and planets travel through here in their circuits of the sky. About every other year Mars makes a visit to Scorpius. So alike in color and in brightness are star and planet that at first glance only someone who knows the background stars can be certain which is Mars and which is Antares.

Scattered all about here are sparkling sights for your binoculars. You should always use them on a colored star like Antares, for the brighter you can make the image of a star, the truer and more brilliant will its color appear. Halfway between Antares and Beta is the rather faint globular cluster M80. Another even better globular cluster, M4, lies two degrees directly west of Antares. The star Zeta lies in a spectacular field of bright companions, including a small misty cluster of fainter stars. In the tail of the Scorpion are two open clusters, M6 and M7; they can be seen with your naked eye and are a gorgeous sight in binoculars.

On sparkling moonless nights when there is no haze above your southern horizon, sweep all through Scorpius and lower Ophiuchus, or here are some of the richest star fields of the Milky Way. When the seeing is just right you will come upon bright star clouds and dark opaque patches that take on added splendor in binoculars. Also catch with your binoculars the double stars Mu and Nu.

LIBRA

Just west of Scorpius is the faint constellation Libra, the Scales. I have postponed mention of this figure until we had the Scorpion to help point it out, for Libra has no star brighter than 3rd magnitude. However, this small, square group is not entirely without distinction: It is the only constellation in the zodiac that represents an inanimate subject. Its name is really out of place since the word zodiac means "Pathway of the Animals." The star Alpha is a wide and easy double in binoculars, and Beta is said to be the only naked-eye star that is greenish in color.

SAGITTARIUS

East of Scorpius and just out of reach of its venomous sting is Sagittarius, the Archer. In mythology this figure sometimes represented a centaur—half horse and half man—with a bow and arrow aimed at Orion, who is now completely out of sight. You can see the bow formed by the curved line of the stars Lambda, Delta, and Epsilon; the tip of the arrow is Gamma. The ecliptic, and therefore also the zodiac, passes entirely across the constellation. At the beginning of winter around December 22, the Sun occupies the position on the ecliptic that is marked X on the key chart. From that point on, it will slowly move northeastward to the spot in Gemini where it stood at the beginning of summer.

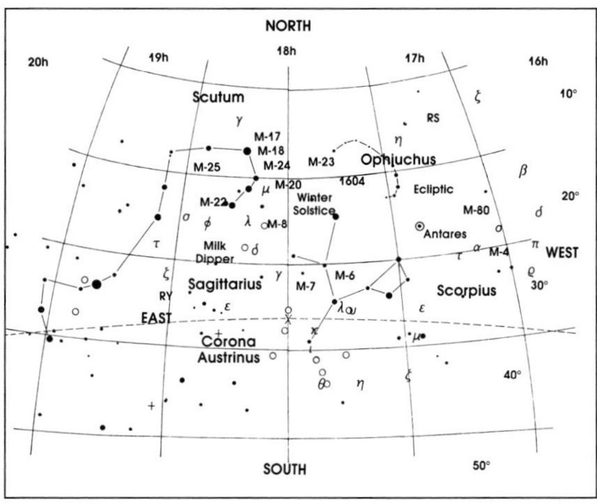

The Region of Sagittarius and Scorpius. Note the abundance of messier nebulae and star clusters. The X on the ecliptic marks the location of the Sun at the beginning of winter.

Sagittarius has a wealth of outstanding objects to challenge your binoculars and observing skill. So closely grouped are many of these points of interest that I have included a detailed chart of this region to assist you in locating them. Near the western border, note M8; this is a star cluster embedded within a hazy cloud called the Lagoon Nebula, and is visible to the naked eye. Just north of it is M20, the Trifid

Nebula. This interesting object can be seen in binoculars, though it takes a telescope to bring out its three-part form. In this same region, M17, M22, and M24 are all worth careful study.

In thickly populated areas such as these, explore the entire region—not just the few special objects I have cited. Sagittarius lies in perhaps the most exciting sector of the entire sky. Here the Milky Way is at its brightest and broadest. Here are the finest star clouds and dark nebulae to be seen from this latitude. And here the densest concentrations of stars anywhere about the skies are to be found. When you turn your binoculars on Sagittarius you are probing deeply into the very anatomy of our Galaxy, as I will explain in the chapter devoted to the Milky Way.

RY Sagittarii is an important variable star you should locate and add to your observing list. It is the same irregular type as R Coronae Borealis and, like it, is usually within easy reach of binoculars. But it too can fade from sight quite suddenly.

LYRA

High in the sky on summer nights is blue-white Vega in the constellation Lyra, the Lyre. It is the brightest of the summer stars. You need no pointer stars to guide you to it for Vega's color and her five nearby companions make identification easy. Vega is a youthful star, which is fortunate, for this makes us even more certain that she will keep her appointment to be Earth's pole star in a mere 12,000 years from now.

The first star to the east of Vega is the "double-double" Epsilon. On a clear night, your naked eye can divide it. My old atlas shows it as two stars side by side, both magnitude 4.5. Binoculars will widen the gap between them but will not, as a telescope will, again divide each of these two stars into a pair. Zeta is also double in binoculars, if they are steadily held on some support. Beta is a variable star ranging between magnitude 3.3 and 4.3 approximately every 13 days. The spectroscope shows that Beta is a binary system with its two components nearly in contact. Their strong mutual gravitation has pulled them into egg-shaped rather than spherical forms. You can follow these variations yourself using Gamma (magnitude 3.2) and Kappa (4.3) as comparison stars.

CYGNUS

Following directly after Lyra, its neighbor to the east is Cygnus, the Swan. More familiarly known as the Northern Cross, you can easily trace out the two axes or arms of this figure. The long one lies in the Milky Way with 1st-magnitude Deneb at the northern end and 3rd-magnitude Beta or Albireo at the southern foot of the Cross, where it represents the head of the fabled swan as it flies south. Albireo is one of the finest double stars in the sky; its components shine at magnitude 3.1 and 5.1 with colors of contrasting yellow and blue. It is believed to be a binary system of extremely long period.

The star marked 61 is a double of magnitudes 5.2 and 6.0. At 11.1 light-years away, it is one of our nearer neighbors and was the first star whose distance was directly measured. Omicron (o) is a wide double whose magnitudes are 4.0 and 5.0 and whose colors are yellow and blue.

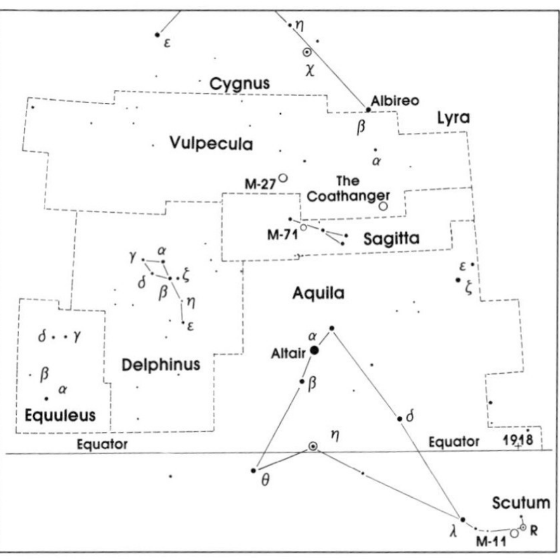

Detail of a group of smaller constellations

The northern Milky Way is at its finest in Cygnus where the region around Gamma, at the crossing of the arms, glitters with a treasury of gems for your binoculars. Find the Northern Coal Sack near Deneb on a night of near-perfect seeing; it is an irregular cloud of opaque dust that blots out the brighter stardust of the Milky Way behind it. The stars you see scattered all about lie between you and the Coal Sack.

Use your glasses to search for the variable star marked with the letter Chi (c) in the long axis of the Cross. You may be able to follow it for about three months of its cycle, as it sometimes brightens to 4th magnitude.

AQUILA

If you are a bird watcher your heart should now feel all aglow, for I have just pointed out to you Cygnus, the Swan, and now comes Aquila, the Eagle. Altair, the bright star of this celestial raptor, has as habitat the mainstream of the Milky Way. It has a star on either side to help with identification. Just a little way below this trio, the star Eta is an interesting variable star that can be followed throughout its entire range. In Chapter 10, I give comparison star magnitudes for this Cepheid-type variable. Watch it carefully and make your own determination of its period. Note where a bright nova suddenly appeared in 1918. For a night or two it equaled Sirius, the brightest of the stars. Today its light is steady at magnitude 11.

Altair is the third point of the celestial figure called the Summer Triangle. Together with Vega and Deneb to the north, this bright lst-magnitude landmark of the skies will be with us all through our summer and autumn evenings. Because Altair is farther south than either Vega or

Deneb, it is the last of the three to arrive on the scene and also the first to leave. Watch for it near midnight in the east on a mid-April evening and bid it au revoir at early dark in January.

SCUTUM

This tiny figure is another invention of Hevelius. Scutum, the Shield, lies in an area of the Milky Way that is sprinkled with luminous star clouds, the brightest being the open star cluster M11, the Wild Duck Cluster. It can easily be detected with the naked eye and makes a fine appearance in binoculars. At a distance of 5,600 light-years, it belongs to our own Galaxy. This entire region is well worth exploring with your binoculars. The variable R Scuti is an important star because of its irregular behavior. It is always within the powers of your binoculars, and you will find it interesting to keep a record of its changes. It also shows the variable Eta Aquilae.

SAGITTA

Halfway between Altair and Albireo at the foot of the Northern Cross you will find Sagitta, the Arrow, in a fine region of the Milky Way. The rather faint cluster M71 lies on the shaft of this realistic looking arrow formed by the constellation's four brightest stars. Ardent bird watchers may prefer to disregard this figure, for it represents the arrow that Hercules has just shot at the two birds, Cygnus and Aquila. I'm sure he missed them both, for the arrow, as we see it, is passing harmlessly between our feathered friends of the Milky flyway.

VULPECULA

Lying between Cygnus to the north and Sagitta to the south is a long, narrow area dimly lit by a few faint stars. Once again Hevelius has been here before us. It seems he had a creative urge, for here he has taken a scattered handful of 5th-magnitude stars and made them into a long, slender fox that carries in his jaws a captured goose—the third bird in this congested aviary of the skies. This constellation, Vulpecula, the Fox, is noteworthy as the home of M27, the famous Dumbbell Nebula (only a misty spot in our binoculars), and for the number of novae that have been discovered within its borders.

Just north of the Vulpecula-Sagitta border you will find a remarkable assemblage of 6th- and 7th-magnitude stars. This is popularly known as "The Coathanger," and even though it appears upside down in your binoculars, you can see at once its likeness to that utilitarian contrivance. This resemblance was briefly disturbed in 1975, when a 6th-magnitude nova suddenly shone out beside the long bar of the hanger.

DELPHINUS

Directly east of Aquila your eye is drawn to the bright, compact figure of Delphinus, the Dolphin. This is a diamond-shaped group of 3rd- and 4th-magnitude stars that is also known as Job's Coffin. The star Gamma is a double of magnitudes 4.5 and 5.5 at a distance of 12", which I find too close to divide with 7x.

Delphinus is the brightest of a confusing group of little constellations without a single star of any note. The chart of this area shows the boundaries of Delphinus, Sagitta, and Vulpecula as related to Cygnus and Aquila. I have also added Equuleus, the Little Horse, which was tethered in this barren spot by Ptolemy of Alexandria some 2,000 years ago.

CAPRICORNUS

A line pointing southeast through Altair and his two companions points directly to Capricornus, the Sea-Goat. To the eye, this figure resembles the outline of an inverted cocked hat of Revolutionary War vintage. It is a constellation for late summer nights. Capricornus never rises high above the southern horizon, so only in the best of seeing will its stars be sharp and steady.

If you found Mizar and Alcor in the Big Dipper to be no challenge for the naked eye, now try Alpha in Capricornus. A clear night will show it to be a double of magnitudes 3.6 and 4.3 lying side by side. With steadied binoculars you may be able to again divide each of these two stars, though the companions are 9th magnitude and difficult. The two bright stars of Alpha are not physically connected by gravitation; their independent motions through space are slowly increasing their distance from each other. Not until the 17th century could these stars be separated by the naked eye. Beta is an easy double. The brighter star is orange and the fainter companion is blue in your binoculars. The colors of the pair remind me of Albireo in Cygnus, though here the distance between the stars is much greater. A third easy double is the faint star Omicron. With your binoculars locate the 5th-magnitude star marked 41. Beside it to the west is the faint globular cluster M30.

Capricornus lies in the zodiac, so all the major planets pass through here in their trips around the Sun. One of them, in fact, was captured here. The planet Neptune was discovered near Delta in 1846.

8

The Autumn Stars

FACING NORTH

In the western sector of the sky at the beginning of this season, the three stars of the Summer Triangle—Vega, Deneb, and Altair—are still in good position for observing while, lower in the twilight, Arcturus leaves our evening skies in mid-September. Above the skyline in the west, you still can watch the strange variable R Coronae Borealis for any sudden changes, and if you are vigilant, you may also find its brother, irregular RY Sagittarii, low in the southwestern sky.

The Milky Way, which in spring lay so low along the northern horizon that it sometimes drowned in haze, now stretches high across the northern sky from east to west and can be observed with ease. With the advent of the longer nights of autumn, it is Ursa Major, the Great Bear, that now becomes entangled in the maze of mist and treetops in the north.

Unless the night is one of sparkling clearness, you may have trouble making out all seven stars of the Big Dipper. Just a glimpse of them, however, will serve to guide you on. A line drawn by your eye from the middle of the Dipper figure through Polaris then onward an equal distance will bring you to a compact group of five bright stars known as Cassiopeia.

CASSIOPEIA

According to legend, Cassiopeia was the queen of Ethiopia and the wife of Cepheus, who stands beside her to the west, quite dull by comparison. At this time of year, Cassiopeia forms a letter "M" in the sky, though as time goes by and season follows season this "M" slowly changes to next spring's letter, "W." But it will still be just across Polaris from the Dipper.

Alpha is a double star that requires a sharp eye and a steady pair of binoculars to separate, since the small blue companion is only 9th magnitude and is virtually flooded by light from the brighter star. The distance between them is some 60", which would be far enough apart to distinguish the two stars in binoculars if they were of equal brightness. Gamma, in the center of the "M," is a remarkable variable star of the most erratic type. The cross marked 1572 shows the location, in that year, of the brightest nova ever recorded. It became brighter than the planet Venus and, for a time, could be seen in full daylight. It is known as Tycho's Star because the great astronomer Tycho Brahe kept an accurate record of its changes for seventeen months until it became invisible to the naked eye.

Cassiopeia abounds in scattered groups and clusters of stars of which M52 and NGC 663 are conspicuous. The region provides interesting sweeping because the Milky Way passes through it.

PERSEUS

Moving eastward from Cassiopeia, you will at once encounter the martial figure of Perseus, the Hero. This bright and beautiful constellation doesn't resemble anything of a familiar shape. It is simply an assortment of curving lines of stars that seem to come together at the star Delta, near the center of the arrangement. The region that lies beneath Alpha is one of surprises and delights. Sweep your binoculars carefully through this sector to discover for yourself the many fields of visual charm and wonder that await you here.

Make believe you are Galileo with the world's first celestial spyglass. He had no chart or guidebook to the skies, but when he turned his homemade scope on a hazy speck of light that he could glimpse in northern Perseus, it was transformed into a maze of tiny, twinkling stars. This famous Double Cluster, which bears the designations h and Chi (χ), can easily be seen on moonless nights making a triangle with the Delta and Epsilon stars of Cassiopeia.

Look at Beta Persei. This is Algol, the Demon Star, and the constellation's most noted attraction. This star is the prototype of a whole class of variable stars. If you watch Algol faithfully from night to night, you will find some point in your vigil when its light suddenly begins to fade. It drops steadily for some 3½ hours to magnitude 3.4 and remains at that level for 18 minutes; it then takes another 3½ hours to get back to its original magnitude of 2.2—nearly as bright as Polaris. Algol remains bright for 2 days, 20 hours, 48 minutes, and 55 seconds, and then the cycle repeats.

Strictly speaking, Algol is not a variable star at all, although the light we receive from it certainly does vary. It all depends on your viewpoint. If the Earth were located in some other part of the universe, Algol's light would burn for us as constant as the Sun's. It is, instead, an eclipsing binary, a 2nd-magnitude star with a much fainter and somewhat larger

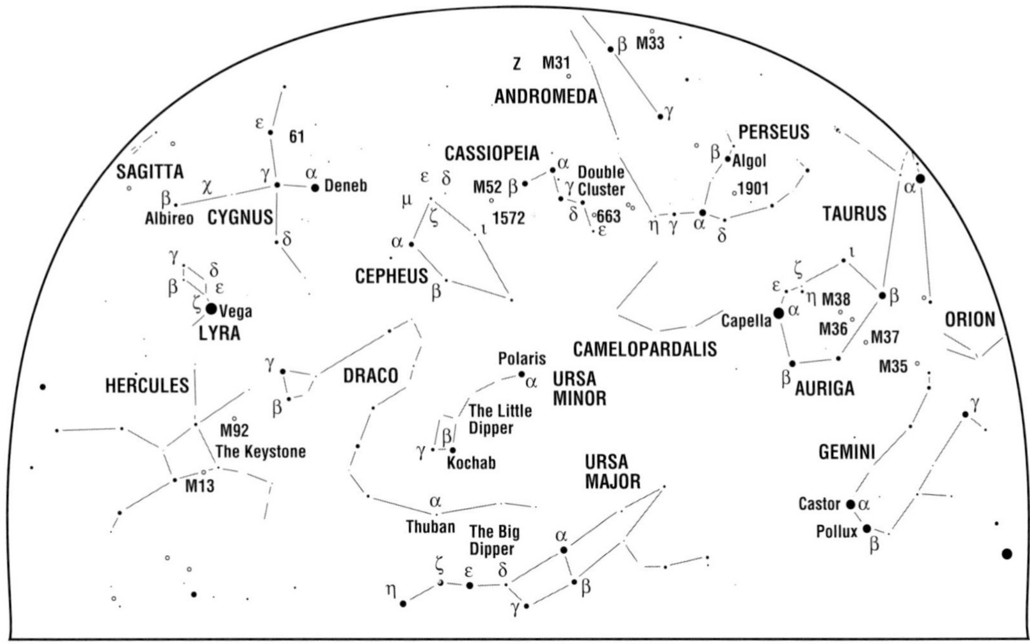

Constellations of the Northern Autumn Sky
Auriga Cassiopeia Cepheus Draco Perseus Ursa Minor

companion star nearby. Together the two stars revolve about their common center of gravity in the period named above. However, the plane of their rotation is tilted somewhat to the Earth so that, from where we stand, neither of the stars ever totally eclipses the other.

When these facts are known, the Demon Star is robbed of its dark mystery. Having watched one eclipse and timed it, you can, knowing its precise period, make your own schedule of future eclipses.

Just above Algol, the open cluster M34 is easily found; it has a nebulous appearance in binoculars. Near the Double Cluster lies the radiant point from which the Perseid meteors seem to appear each year on or about August 12. In February 1901, a nova about as bright as Vega or Arcturus suddenly appeared at the spot marked with a cross. Today, it is still a restless star of magnitude 13.

AURIGA

Continuing eastward along the broad pathway studded with the stars of Cassiopeia and Perseus, we come to the pentagon formed by the five brighter stars of Auriga, the Charioteer. Capella, a star as bright as either Vega or Arcturus, is the undisputed ruler of this region and the most northerly of all the lst-magnitude stars. You may note that Auriga seems

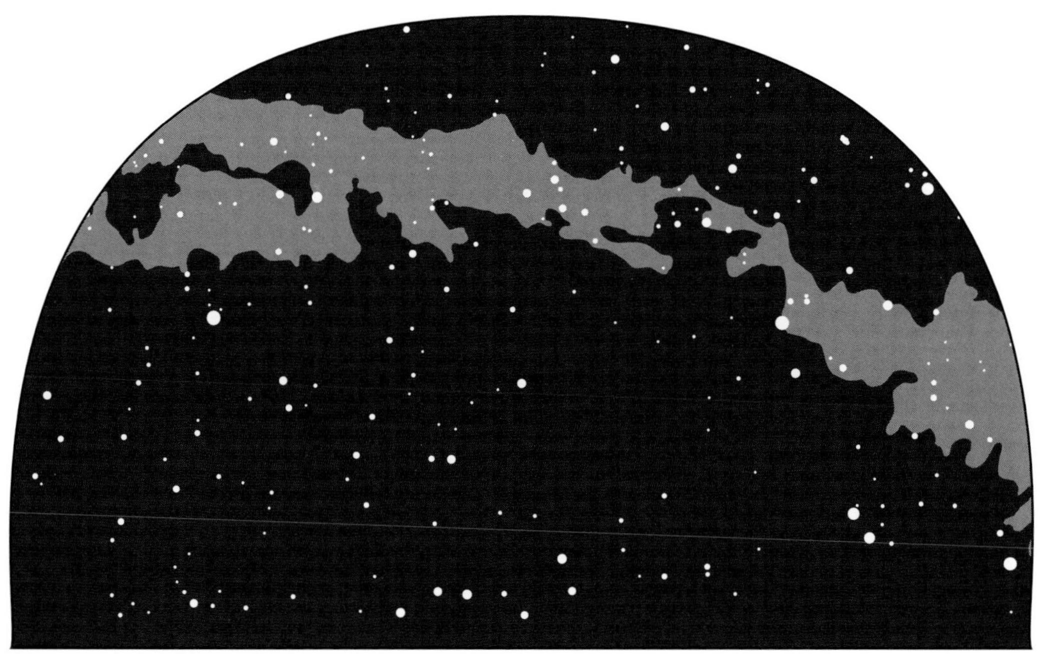

The Northern Autum Sky As Seen On:
September 15 at 12 a.m. *October 1 at 11 p.m.* *October 15 at 10 p.m.* *November 1 at 9 p.m.* *November 15 at 8 p.m.* *December 1 at 7 p.m.*

to have two stars named Beta—one to the north, the other in the south. Actually, the southern Beta marks the tip of one of the horns of Taurus, the Bull, though it is also usually included in the pentagon.

Near Capella, you will find a small acute triangle of 3rd- and 4th-magnitude stars often called The Kids. Epsilon, at the apex of the triangle, is a most remarkable star. Like Algol in Perseus, it is an eclipsing binary, but where the period of Algol is about 2½ days, that of Epsilon and its dimmer companion is 27 years! Zeta, the fainter of the stars in the base of the triangle, is still another eclipsing binary, with a period of a little more than 2½ years.

The Kids provide a good means of identifying Auriga and Capella at any time or season and at any position in the sky. In the lower portion of the constellation you will find rich sweeping for your binoculars, for the main stream of the Milky Way flows through here. You can readily discover such starry islands as the open clusters M36, M37, and M38.

FACING SOUTH

Just as the migrating bobolink and flicker bring back to our doorstop the chickadee and junco, so too the dropping in the west of the starry figures of summer brings up in the east a host of visual delights to keep our eyes pointed skyward through all the nights of autumn. High up in the east you'll find a horse with wings, a maiden bound in chains, and a neighboring universe. In deep mid-skies you'll sight a monstrous whale and, far to the south, a lonely fish with a single bright star in its mouth.

THE SOUTHERN FISH AND FOMALHAUT

Just a little west of due south on a mid-autumn evening, you will find 1st-magnitude Fomalhaut in the constellation Piscis Austrinus, the Southern Fish. Fomalhaut is the most southern of all the 1st-magnitude stars visible from our latitude. It is almost exactly the same brightness as Pollux, a part of Gemini, the Twins, which you saw in the winter months and which already may be rising low in your eastern sky. There is nothing about this Southern Fish that suggests one of the finny tribe, but Fomalhaut does make one corner of a small, four-sided figure. Since it is the only bright star in this region of the sky, identification is easy. The star Beta, in the opposite corner of the quadrilateral, is a double star, though horizon haze may make its 8th-magnitude companion difficult to see in low-power binoculars.

Fomalhaut rises low in the southeastern sky at almost the same instant that Capella appears above the northeastern horizon. However, Fomalhaut is so far south that it makes its brief march across the sky and sets long before Capella reaches the meridian, its half-way mark.

PEGASUS

Directly above Fomalhaut you will find a characteristic figure of the autumn nights—the Square of Pegasus. Here are four stars of almost 2nd-magnitude brightness making a square of nearly equal sides. The two stars that form the west side of the square point directly downward to Fomalhaut. The brightest of the Square stars is Alpheratz in the upper left-hand corner, though it does not really belong to Pegasus but to Andromeda, as we shall see later. This is another case of star-borrowing, such as was done with Beta from the horn of Taurus in order to complete the pentagon of Auriga.

Pegasus, the Winged Horse, lies in a sparsely populated region of sky, and for this reason not more than a figurative handful of stars can be seen with the naked eye within the large area enclosed by the Square. The globular cluster M15 is easily seen with binoculars, though the individual stars that compose it cannot be resolved without a telescope.

AQUARIUS

Just east of Capricornus and north of Fomalhaut lies the large, dull, shapeless constellation Aquarius, the Water-Bearer. The fact that it is a zodiacal constellation gives it some added importance. It was near the star Delta that an unknown planet was once seen but not recognized. In 1756, Tobias Mayer almost discovered Uranus here, but it was moving so slowly that he recorded it as a fixed star. William Herschel found it again, but not until it had crept eastward into Gemini twenty-five years later. And Herschel at first mistook it for a comet.

Note the four-starred figure called the Water Jar and, to the east, the streams of small stars that represent the water pouring from its mouth. The globular cluster M2 is a good subject for binoculars, though you must look sharp because it is a compact cluster that can be mistaken for a star. On the best of nights you can faintly see it with the naked eye. With your binoculars sweep the entire region. The peculiar variable star R Aquarii, unfortunately, is generally too faint for our binoculars. About a degree and a half west of the star Nu lies the Saturn Nebula, a gaseous nebula that your binoculars may show as just a speck of haze. On the key chart it is numbered NGC 7009.

PISCES

Two faint streams of stars border the Square of Pegasus on the south and east. With the naked eye only the Circlet, which represents the open mouth of the Western Fish, is bright enough to be easily made out. With binoculars both the Western and the Northern Fish can readily be traced, though in the Northern Fish Eta and Zeta lie in Andromeda's territory. At the spot marked "X," the Sun crosses the equator heading northward on the first day of spring. The ecliptic, the equator, and a line passing through both north and south celestial poles all intersect at this point. This north-south line is the zero hour circle, and all star positions are measured east from it. It is the Greenwich Meridian of the sky.

In my old sky atlas, these two fish are tied together with a long ribbon attached to their tails. Two winding streams of faint stars represent this ribbon, which loops around Alpha, the constellation's brightest star.

CETUS

If you follow the faint star stream that ties the Western Fish to Alpha, it will lead you almost to the head of Cetus, the Whale. Appropriately, this is one of the largest of all constellations in area, though there is nothing about it to attract your casual attention. Beta is its brightest star at 2nd magnitude. The head of the monster is marked with 3rd- and 4th-magnitude stars in the shape of a small pentagon, which you will soon learn to recognize.

Cetus's chief claim to fame is a star you may not find at all on your first search. It is Mira, or Omicron, the first variable star ever discovered. It

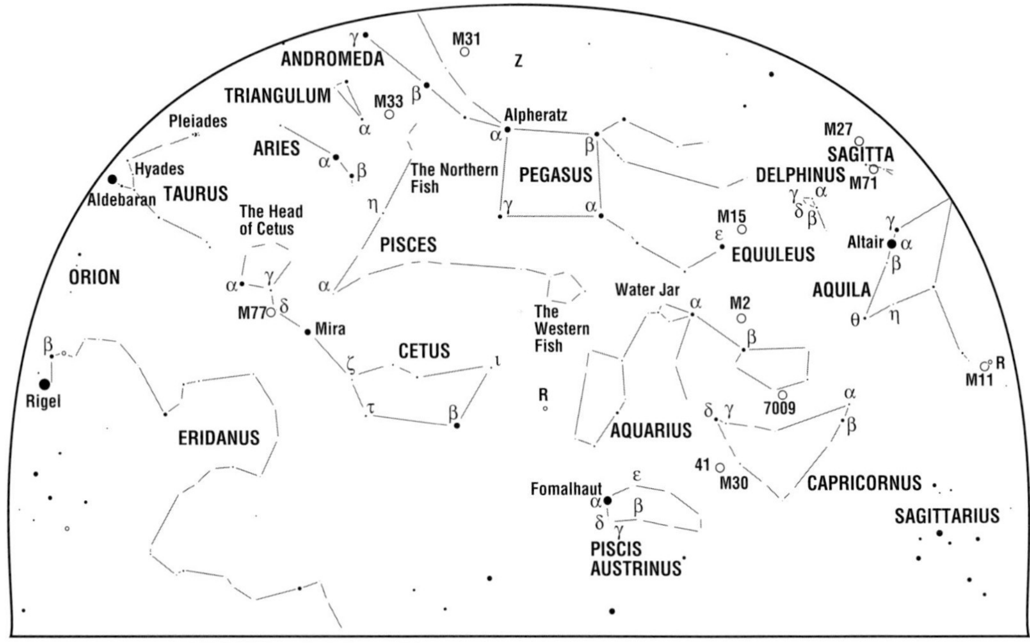

Constellations of the Southern Autumn Sky
Aquarius Aries Andromeda Cetus Pegasus Pisces
Piscis Austrinus Triangulum

was found in 1596 by the astronomer David Fabricius, who noted that it was 3rd magnitude in August of that year. By autumn it had disappeared.

The star marked 66 in the field near Mira is a close double in binoculars. Alpha is a wide double in binoculars; its magnitudes are 2.5 and 5.5.

Closely following the star Delta to the south, look for the faint spiral galaxy M77. It is an elongated, misty spot whose dim light has traveled 50 million years to reach us.

ANDROMEDA

Finding this constellation is a simple matter. The northeastern star in the Square of Pegasus is also the Alpha star of Andromeda. From this starting point the constellation figure sweeps to the left in two long, curving arcs. The lower arc contains three 2nd-magnitude stars that point almost directly to 2nd-magnitude Alpha, or Mirfak, in Perseus.

The outstanding attraction in Andromeda is M31, the Andromeda Galaxy. You can easily find it with your naked eye on a good night, and binoculars make it a brighter, hazier patch of light that over the years has often been reported as a comet. At a distance of 2.2 million light-years, it is the most distant object the unaided human eye can see. In 1885, a 6th-magnitude supernova suddenly appeared in M31.

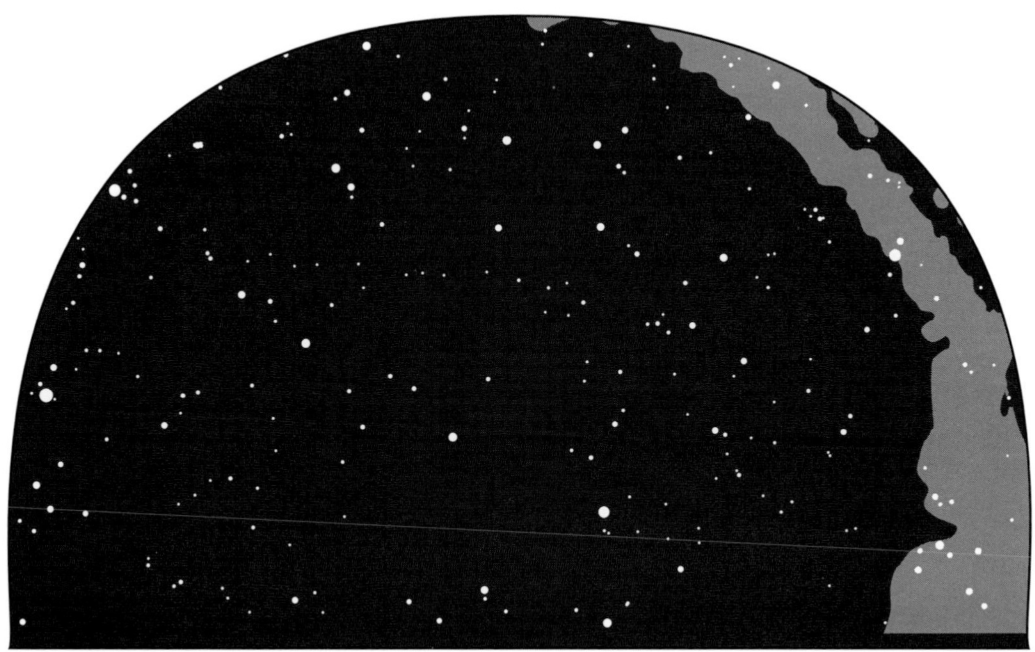

The Southern Autumn Sky As Seen On:
September 15 at 12 a.m. *October 15 at 10 p.m.* *November 15 at 8 p.m.*
October 1 at 11 p.m. *November 1 at 9 p.m.* *December 1 at 7 p.m.*

Several of autumn's best constellations are prominent in mythology. Two of these are King Cepheus and his wife, Queen Cassiopeia, both of whom, you'll remember, we have already met in our exploration of the northern skies. When the Queen boasted that she was fairer than the sea nymphs, the wrath of Neptune, god of the seas, was aroused. He decreed that Andromeda, the beauteous daughter of the royal pair, must be sacrificed. She was chained to a rock beside the sea and left to be devoured by Cetus, the terrible monster of the deep.

Fortunately, however, our hero Perseus mounted the winged horse Pegasus and dashed to the rescue of the maiden. Mythology further tells us that Andromeda rewarded Perseus for his valor by becoming his wife. They must have lived happily ever after, for as you can see, they still are sparkling side by side. And so this little drama of the skies ends on a happy note, a fitting final curtain for a truly all-star cast.

TRIANGULUM

Beneath a line connecting the Beta and Gamma stars of Andromeda is a small constellation called Triangulum, the Triangle. Its stars are of 3rd and 4th magnitude. With your binoculars look for M33 to the right of the sharp point of the triangle. Like M31 in Andromeda, this is also a galaxy.

I have often glimpsed M33 with the naked eye, but only when it was high up in the sky, well above the haze of the horizon. You must have a clear, dark night and a location free from all interfering lights. With binoculars you will find it much fainter than M31. Though M33 is at its visual best in low-power binoculars, you will get no hint of its marvelous spiral pinwheel structure so striking in long-exposure photographs.

Ceres, largest of the asteroids and the first to be discovered, was found in Triangulum on January 1, 1801—the first night of the new century.

ARIES

Here is another triangle, though this one is obtuse. It lies between Triangulum and the head of Cetus, where its only conspicuous figure is made up of three stars of 2nd, 3rd, and 4th magnitudes. Aries, the Ram, has little of observational interest for us unless it has planetary guests. Because Aries is a zodiacal constellation, the Moon and planets pass through here to get from Pisces into Taurus. Two thousand years ago, the point where the Sun crossed the equator on the first day of spring lay in Aries, though now, as we have seen, it lies in Pisces. Aries is still the first sign in the zodiac of the astrologer and the old-time fortune teller, even though the Sun, because of that peculiar wobble in our polar axis, has long since postponed its appointment with the Ram and won't arrive here for another month.

I have remarked before that as each season in the skies gives way to the next, the advance scouts of the approaching company are already well in sight. Even as you watched small Aries on autumn nights, you must have seen, just a little to the east, a tiny, misty group of stars, the Pleiades, that should now be old acquaintances, for they danced their way all across our winter skies. Just as the chilly days of fall bring returning juncos and red-polls to our feeders, the crisp autumn nights bring to the eastern sky the baleful eye of Taurus. When, on a clear night in November, I hear the five notes of a great horned owl coming from the shelter of our pines, I also see the giant figure of Orion standing just above the treetops as though eager to begin his nightly march across our winter skies.

9
The Milky Way

In the sky map drawings, you saw the staggered path of the Milky Way across the skies of the seasons of the year. In the winter skies it is a rather narrow band that rises upright from the northern horizon, passes nearly overhead, then, like a river lighted by bright beacons flanking either shore, flows southward until it drops behind the Earth just below the great star Sirius.

In the evening skies of spring, the Milky Way shows only in the north, though you must have a sparkling night and a clean sweep all along your northern horizon if you want to find it. Rely on your binoculars to brighten up this strip of sky, for the "milkiness" is overwhelmed by mist. The course of this dim trail is best marked by the constellations through which it passes. From west to east, Hiawatha's "pathway of the ghosts" skirts the northern skyline from southern Auriga through Perseus, Cassiopeia, Cepheus, and the Northern Cross.

In summer, and particularly in the month of August, the Milky Way is at its best. This is the season of the year when, on a sparkling, moonless night, you should arm yourself with your binoculars and make a special expedition to some favorite country spot with a clear horizon all around and no distracting lights. There you will see the Milky Way in all its astonishing glory.

Many years ago I was the guest of an observing group that regularly drove some forty miles to a lonely desert hilltop to carry out their program of estimating variables, plotting meteor trails, and photographing deep-sky objects. While my companions were setting up their instruments, I walked alone to an isolated rise a couple of hundred yards away and looked about me. From that arid, treeless spot I could just make out the far horizon as a dim, unbroken circle where the darkness of the Earth gave way to sparkling stars. Below and all around there was only empty blackness, but across the sky above me the Milky Way, brighter by far than I had ever seen it, was like a long, slack line supported in the zenith by the Northern Cross and hung from end to end with all the trailing, filmy garments of the night that might just have been washed and hung up high above the Earth to dry.

Soon my friends came searching for me and I returned with them to their observing site. But the eerie setting—the utter silence of that lonely desert hill, the velvet blackness of the night, and over all the bright course of the Milky Way with its pools of murky shadow and stepping-stones of starclouds—made that night an occasion that I still hope will sometime come again.

Obviously, such lucid nights can never happen in most people's environments. The world is too much with most of us. But we can try for something like it when we travel or are on vacation to a unspoiled and uncrowded area.

Let's assume you have found a good location for observing and that it's a clear, moonless summer night. Depending on the evening hour, some of the brighter stars of Perseus should now be showing in the low northeast. You have already swept this region with binoculars in the autumn season and will recall the brilliant streams of stars your binoculars revealed near Mirfak, the constellation's brightest star. Above it lies the famous Double Cluster and the five bright stars that form the "W" of Cassiopeia. Still higher in the sky, the Milky Way flows by the foundation of the "house" of Cepheus, and right here you can pause again to note the brightness of the variable Delta Cephei and the redness of Mu, the Garnet Star. This brings you to the borders of Cygnus and the Northern Cross. Here the whole character of the Milky Way seems to change abruptly. Throughout this entire northern sector from Perseus to Cygnus, the Milky Way has been a single, rather narrow trail liberally sprinkled with a host of distinct, sparkling stars. But it hasn't been a dusty path. There has been little of the hazy or "milky" appearance you will encounter from this point southward.

Two large, dark areas bridged by a brighter spot can be seen near Deneb at the top of the Northern Cross. This region is sometimes called the Northern Coal Sack. Halfway down the long arm of this Cross, just south of the star Gamma, you will find the beginning of the long, dark lane known as the Great Rift, which for a distance divides the Milky Way into two narrow streams. At the start, the western branch is the brighter of the two, but this soon fades and completely disappears in the region of the tail of Serpens. Meanwhile, across the Rift, the eastern branch of the divided stream steadily increases in brightness through Sagitta and Aquila, with Altair and its companion stars glowing near the center of this branch. You will find one of the brightest spots of all in the tiny constellation Scutum. This is the open cluster Mll with its bright star in the center. Just to the west of it, you will want to pause again to jot down your estimate of the perplexing variable star R Scuti, which is always well within the reach of your binoculars.

South of Scutum, the Great Rift leads into Sagittarius. Here, like sunlit islands in a stream, are some of the broadest star clouds of the entire Milky Way. Sweep carefully over the region below the handle of the inverted Milk Dipper. M20 and M8 are outstanding and can be seen without binoculars. In lower Sagittarius, the Great Rift ends and the Milky Way attains its broadest width, extending well into Scorpius to include red Antares among its sparkling fields. Search the region about two degrees above Antares where 5th-magnitude Rho (ρ) Ophiuchi is the center of a dark nebula. These seemingly blank areas are not holes in the starry fabric of the Milky Way but are dark clouds of cosmic dust obscuring the light of the myriad stars that shine unseen behind them. The Coal Sack in Cygnus, as well as the long line of the Great Rift, are fashioned out of this same non-luminous material.

You will note that this southern section of the Milky Way, particularly in Sagittarius and Scorpius, is much brighter and has much more diffuse haze than the northern path. This is why the Milky Way of autumn is less spectacular than that of summer, for in early fall this bright, broad region is setting in the west and is lost to sight. So, while summertime and Sagittarius are still with us, let's try to grasp the real significance of this "broad and ample road whose dust is gold and pavement stars."

Either with the eye or binoculars, it's difficult to comprehend the appearance and structure of the Milky Way since you are watching from a point that lies well within the Milky Way. But you are with noble company, for all the stars you see at night—together with the Sun and his family of nine planets, including Earth—are fellow members of this same great galactic system. In a sense, we also are its prisoners, for only in imagination are we free to leave this tenuous assemblage of stars and cosmic clouds and interstellar dust to view it from outside its borders.

Scattered all about the skies in every direction are millions of other Milky Ways more or less like ours. You have already been introduced to some of them in our autumn skies: M33 in Triangulum and M31 in Andromeda were pointed out as nearby and bright examples. Some galaxies turn their faces directly toward us and others present an edge-on view. We see M33 as though we were looking directly at its center, just as a sparrow hawk might view a clover field as he hovers above it in mid-air. This is possibly what our Milky Way Galaxy would look like to some stargazer watching us from a planet in M33.

Abdul Hadi Marafie

The Andromeda Galaxy is the brightest spiral in the sky.

Inasmuch as the Milky Way is some 100,000 light-years in diameter and about 10,000 light-years thick, our Sun appears to be about 30,000 light-years from the center of the Galaxy and midway between the edges of the galactic disk.

But what proof have we that these positions are correct? With nothing but your eyes and your binoculars, you can see much of the evidence for yourself. Look again at the edge-on view and imagine that you are standing at the crossmark looking toward the bulge in the center. That bulge is the center of our Galaxy, and we saw it in the summer skies when we looked at Sagittarius and remarked how broad and bright the Milky Way was in that region and how thickly the star clouds were heaped about when we swept our binoculars through there. Now, in imagination, turn around and look directly away from the center bulge. You are looking at the winter Milky Way, at such regions as Taurus, Gemini, and Monoceros. Here, you will recall, there were no star clouds and the stars themselves were relatively few and far between compared to Sagittarius and Scorpius in our summer skies.

To verify the Sun's location near the middle of the edge-on view of our Galaxy, select a time when the Milky Way is overhead and divides your sky into equal halves. Now roughly compare the number and distribution of the stars you see. As these halves seem to be about equally populated with stars, it shows that you must be standing near the marked position and are looking toward the regions of the north and south galactic poles. The dark lanes you see crossing through the plane of this edge-on view are the dark areas of the Milky Way, such as the Great Rift, the dark nebulae, and the empty spaces between the spiral arms.

Chris Cook

The summertime Milky Way arches over Texas.

As these spiral arms so strongly suggest, the entire Milky Way is in rapid rotation about its bulging center. This whirling motion gives our Galaxy its flattened disk-like shape. It is believed that our solar system is rotating around the center hub at a speed of about 150 miles per second, which amounts to one complete revolution in about 250 million years.

In looking at the Milky Way through binoculars on a sparkling, moonless night, you will have the feeling that certainly those stars must be crowded close together, almost like a wheeling flight of blackbirds at roosting time in autumn. Since all these stars are in rapid motion it seems inevitable that there would be collisions, for in our binoculars many of them appear to be almost in contact. This, of course, is all a matter of perspective. Each of the 200 billion stars in the Milky Way has its own home territory in a vast space 100,000 light-years across. This would seem to give each one ample room to roam around, for no one has ever witnessed a collision of two stars. Our own Sun, a typical star of the Milky Way, is not exactly crowded since its nearest neighbor star, Proxima Centauri in the southern skies, is some 24 trillion miles away!

Of the hordes of stars that form our Galaxy, the number that we can see with the naked eye seems pitifully small. On clear, sparkling nights, I have sometimes asked visitors how many stars were visible to them. Their guesses have ranged from "thousands" to "millions" to "countless." Actually, the number that can be seen at any one time is about 1,500. On such occasions, I always remember the over-generous star counts of the little child in Robert Louis Stevenson's charming rhyme, "Escape at Bedtime":

The lights from the parlour and kitchen shone out
Through the blinds and the windows and bars;
And high overhead and all moving about,
There were thousands of millions of stars.

There ne'er were such thousands of leaves on a tree,
Nor of people in church or the Park,
As the crowds of the stars that looked down upon me,
And that glittered and winked in the dark.

The Dog, and the Plough, and the Hunter and all,
And the star of the sailor, and Mars,
These shone in the sky, and the pail by the wall
Would be half full of water and stars.

They saw me at last, and they chased me with cries,
And they soon had me packed into bed;
But the glory kept shining and bright in my eyes,
And the stars going round in my head.

The starry skies are also an escape for some who are no longer little children. After an evening of exploring the glories of the Milky Way through binoculars, even a seasoned watcher sometimes may put away the binoculars and retire with the stars circling in his head.

10
Variable Stars

As a stargazer, the celestial specimens that are the most likely to capture and hold your attention and keep you on the edge of your observing chair are those stars whose light output is subject to change or fluctuation. These stellar show-offs are the variable stars.

From time to time earlier in this book I have called attention to certain of these inconstant stars that you could observe throughout all or at least the brighter part of their cycle of change. The charts in this chapter show the naked-eye appearance of the sky field or constellation figure in which each of these stars is located.

As with the seasonal sky maps, I have used connecting lines to make a star pattern more readily recognizable. In nearly all cases, you can then find the variable itself, represented by an encircled dot (⊙) near the center of a large circle that represents the immediate sky field as you will see it through 7x binoculars. This circle is ten degrees wide and includes stars down to magnitude 7.5.

Many of the charts also include a small inset view of the field around the variable. This is an enlargement made to extend the usefulness of the charts for those equipped with binoculars of higher power or small telescopes. On all charts, north is up.

On each chart you'll find the star's designation number, its period in days, and its magnitude range in brightness. The designation number is always a six-digit string that tells the star's location in the sky and helps you plot its position on a star chart or atlas. For example, the designation number of the variable star Delta Cephei in the constellation Cepheus is 222558. The first four figures of this number place Delta Cephei at 22 hours, 25 minutes east of the vernal meridian of the sky. This is the star's right ascension, which corresponds to longitude in geography. The final two figures are the star's declination. They indicate that Delta Cephei is 58

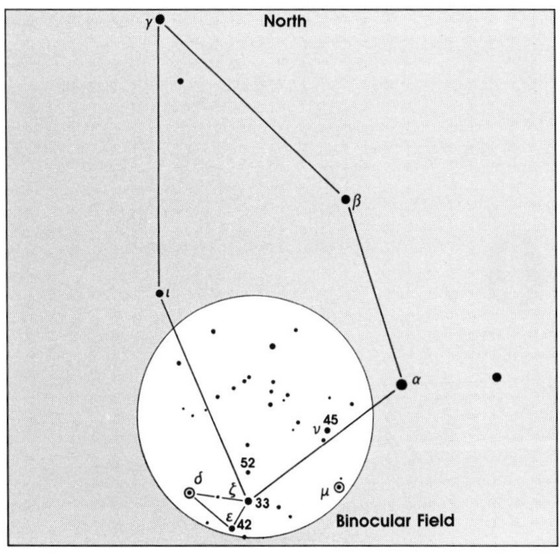

5⅓ Days 222558 Delta (δ) Cephei
Range: 3.5–4.4

degrees north of the celestial equator, an imaginary line passing around the sky directly above the Earth's equator. For stars south of the sky's equator, the two declination figures are underlined. Plot the right ascension and declination coordinates of an object on your atlas, then search in that region of the sky. You can easily become a variable star observer if you take these charts, your binoculars, and a red filtered flashlight out on the first clear night and find one or more of these variable stars for yourself.

In any season or hour of the night you are certain to find some of these fields in good position. You may already have done this by using the naked-eye sky maps of the seasons. If so, do it again, for you are almost certain to find that the star you saw before has now changed in brightness. Stars like Beta Persei and Delta Cephei are naked-eye variables that are always in your northern skies; you can find them with binoculars even on moonlit nights. With some of the others, however, don't be surprised if you locate the proper field only to find the variable missing. Don't be discouraged by its absence. Keep watching that empty spot, and sooner or later you'll catch the first faint glimmer of its growing brightness.

I have drawn these charts with north at the top. This is correct for use with the naked eye and binoculars, though not for most telescopes, which invert images. Hold the chart so its north point is toward Polaris, no matter whether the variable you are hunting for is in the east, south, west, north, or even underneath the pole star. When held this way, the chart and the sky field will be oriented the same way.

Once you have found the variable, it's a simple matter to estimate its brightness by comparison with stars that don't vary. On each chart you'll find an assortment of comparison stars with their magnitudes listed beside them. For example, look at the chart of the variable star 184205 R Scuti on page 72. Near the variable are three comparison stars marked 61, 67, and 71. These are magnitudes figured in tenths; I have omitted the decimal point so you won't mistake the point for a fainter star. The star marked 61 is, therefore, magnitude 6.1, or just visible to the naked eye.

Now compare the brightness of the variable R with these three stars.

(Remember, with star magnitudes, the larger the number, the fainter the star.) In these three stars there is a spread of one full magnitude, from 6.1 to 7.1. Perhaps R looks fainter to you than 61 and brighter than 71. If it appears midway in brightness between the two, then its magnitude is 6.6. If not, then the star 67 will help you pinpoint your estimate.

Many variable stars, such as R Leporis and R Leonis, are decidedly red in color. Because these stars seem to grow brighter if you watch them steadily, estimating their magnitudes should be done by taking quick glances so you won't overestimate their magnitudes.

In general, variable stars have been named in the order of their discovery in the constellation where they are located. A few, such as Algol and Mira, were given proper names long ago; others—for example, Delta Cephei and Eta Aquilae and Chi Cygni—have Greek letter symbols. But most have been given capital letter designations starting with the letter R. Thus R Leonis was the first variable to be found in the constellation Leo; Z would be the ninth. After Z was reached, paired letters were assigned beginning with RR as the tenth variable, then RS to RZ, SS to SZ, until ZZ was reached. After this came AA to AZ, then BB to BZ and so on until QZ was reached. The letter J was always omitted. This letter system accounts for 334 variables per constellation, but for the more prolific constellations, notably Sagittarius and Cygnus, even this wasn't enough. A number system was devised to follow QZ with V335, V336, etc. At latest count, Cygnus is up to V1815 and Sagittarius is up to V4091.

After you have located a variable star and made your estimate, the next step is to keep a permanent record of its magnitude together with the date, hour, and any significant details of the observation. These estimates should be combined with those of other observers where they will be available to any interested researcher at any time. An estimate you have made may just be the only one in the world made of a particular star at a given hour of a certain night.

Once, I was able to furnish estimates of an old nova, 155526 T Coronae Borealis, which I had recorded fifty-five years before. And I still treasure my penciled record of the first estimate I ever made. It was dated March 1, 1918, and was written on the ersatz note paper of that era.

If, in following the stars I have charted here, you begin to feel the first blissful birth pangs of an awakening interest in the systematic observing of variable stars, I urge you to write to:

The American Association of
Variable Star Observers
25 Birch Street
Cambridge, MA 02138

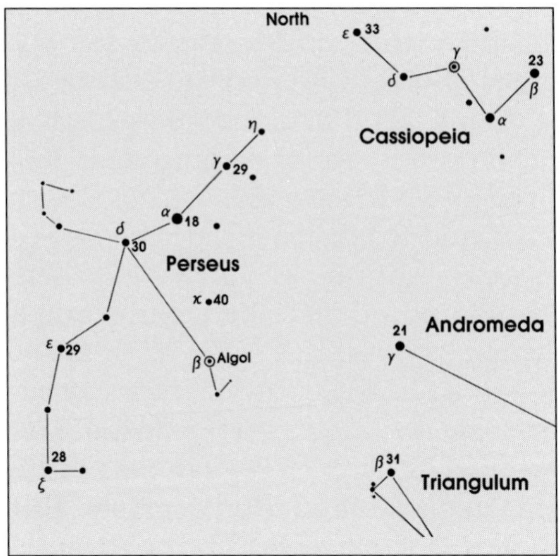

Irregular 005060 Gamma (γ) Cassiopeiae
Range: 1.9–2.8

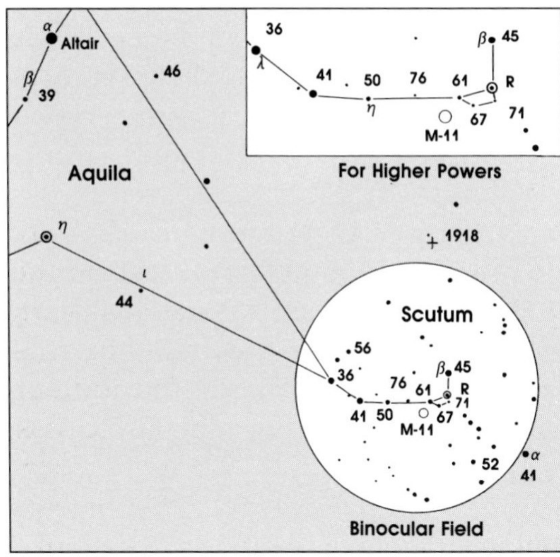

Semi-regular 184205 R Scuti
Range: 4.7–4.8

The director of this worldwide association of observers will be pleased to send new members detailed instructions for making, recording, and submitting observations. The list of newly found and still neglected variables is constantly growing, and there is still much pioneering work awaiting the earnest binocular observer, particularly in estimating and recording the changing brightness of semi-regular, eclipsing, and Cepheid variable stars. Make a member of this last class—a Cepheid variable—your first variable star. Like the Big Dipper, it is always in the sky and you can't fail to find it the first time you try.

DELTA CEPHEI

This circumpolar variable star can be followed with your naked eye throughout its entire range. Binoculars are a great help, however, on moonlit nights or when the circling seasons or the passing hours have moved the "house" of Cepheus into that hazy region between Polaris and the northern skyline. Delta forms the apex of a small triangle of stars at the base of the house. Its changing light can easily be noted by watching it on two or three successive nights. At maximum brightness it equals Zeta, its neighbor in the triangle; at minimum it is slightly fainter than Epsilon, its other neighbor.

Delta Cephei is the prototype of a class of variable stars known as Cepheids. Stars of this class are also called pulsating stars because of the strange way their fluctuations are produced. The stars alternately expand and contract, much as our heart does in its rhythmic beating.

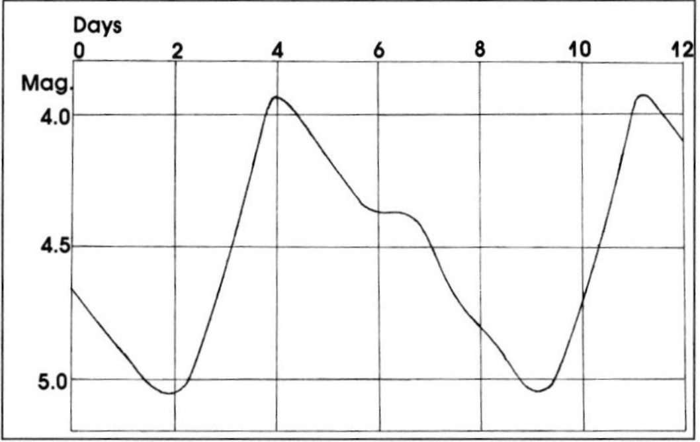

The Light Curve of Eta (η) Aquilae

Such stellar pulse rates are most precise, though by human standards they are quite prolonged: each of Delta's heartbeats of Delta takes exactly 5 days, 8 hours, 47 minutes, 39 seconds. This is its period, or the interval between one maximum and the next. The frequency of these heartbeats has solved the age-old mystery of how far away many stars are. Knowing the period and the apparent brightness of a Cepheid, astronomers can determine the star's luminosity. The difference between the actual luminosity and the observed brightness gives us the distance to the star.

Delta Cephei is 1,000 light-years away, a fairly neighborly distance. On the next clear night make a house call and, with your binoculars as a celestial stethoscope, monitor the heartbeat of Delta Cephei. Cepheids have also been found in star clusters and distant spiral galaxies, where their pulsing throbs have helped to narrow astronomers' estimates of the size and age of the universe.

GAMMA CASSIOPEIAE

This is a most eccentric star, one so unpredictable that no definite period can be assigned to it. In 1937, I watched it brighten to magnitude 1.6, nearly equal to Deneb in the Northern Cross (magnitude 1.3) and, therefore, definitely brighter than nearby Alpha in Perseus. After this surge it slowly dropped to magnitude 3.0. It can usually be found near magnitude 2.5 but should be watched occasionally for any change. It has been suggested that Gamma is a rapidly rotating star that casts off material at intervals.

When estimating Gamma, use comparison stars at about the same altitude above the horizon as the variable so your estimate isn't influenced by any difference in atmospheric density. Don't use nearby Alpha as a comparison star, for it has been suspected of being slightly variable.

The chart also helps you follow the changes in the eclipsing variable Algol in Perseus, already described among the autumn stars.

R SCUTI

This is the star I used earlier to illustrate how to estimate magnitudes. I chose it because it has such an accommodating array of nearby comparison stars. It lies in an easy field to find, for the bright cluster Mll is in the

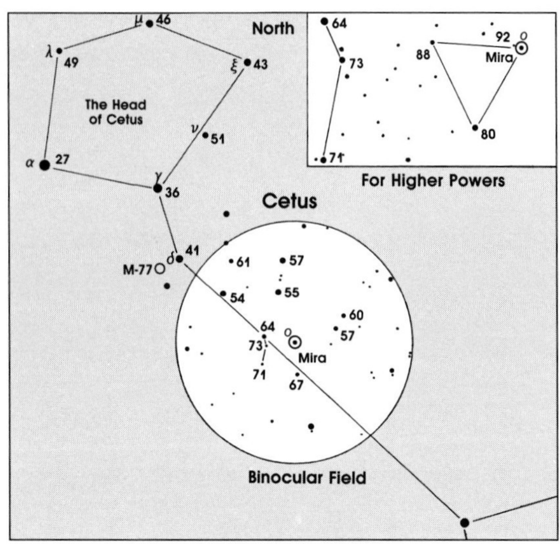

332 Days 0214<u>03</u> Omicron (o) Ceti
Range: 3.7–9.2

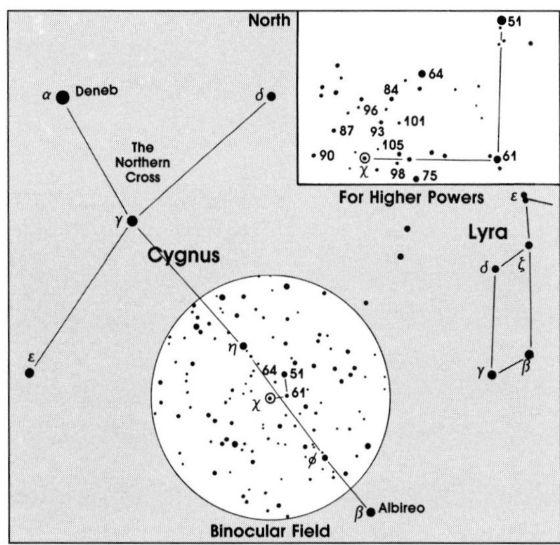

406 Days 194632 Chi (χ) Cygni
Range: 5.1–13.3

same binocular field just to the left of the variable. Discovered in 1795, R was the tenth variable to be found in the entire heavens. It is a reddish star with a somewhat irregular period of 146 days. I have often found it at magnitude 5.0 but never fainter than 8.0. This is a brilliant region of the Milky Way that you'll find especially rich in your binoculars.

ETA AQUILAE

Eta Aquilae is a Cepheid variable very similar to Delta Cephei but with a slightly longer period and a greater range. It is located just a twinkle north of the sky's equator. This star is characterized by a pronounced hump about midway in its descending light curve. If you plot your own light curve or graph of this star's changes, it will show a hump coming about two days after maximum when the fading light holds steady for a time. Eta and its comparison stars are shown on the chart of R Scuti.

MIRA (OMICRON CETI)

Mira is the "senior citizen" of the variable star clan—the first variable star discovered. It was first recorded by the Dutch amateur astronomer David Fabricius in August 1596, when it was at 3rd magnitude, or about equal to the faintest of the Big Dipper stars. A few weeks later it had disappeared. Johann Bayer, who gave the stars their Greek letter designations, saw the star in 1603 and, not knowing of its variable nature, assigned it the letter Omicron (o) in his atlas.

Since Mira has been under observation longer than any other variable, you might suppose that its behavior could be precisely forecast, but this is not the case. Both the date of maximum and the magnitude at maximum are subject to irregularities.

I have never seen Mira quite as bright as its neighbor Beta Ceti, which glows at magnitude 2.0. As I write this, only two days before its predicted date of maximum, I estimate it at magnitude 3.2, which just shows that Mira doesn't go by the book.

Mira is a fine binocular variable. It can usually be followed through most of its range, though when it's faint, a nearby 9.2 star can be confused with it. Like Delta Cephei, Mira heads a class of variables. These are known as Mira or long-period stars because their periods range from 200 to 400 days. Mira is a red giant, similar to Betelgeuse in Orion. It is immense—about 400 times the size of our Sun.

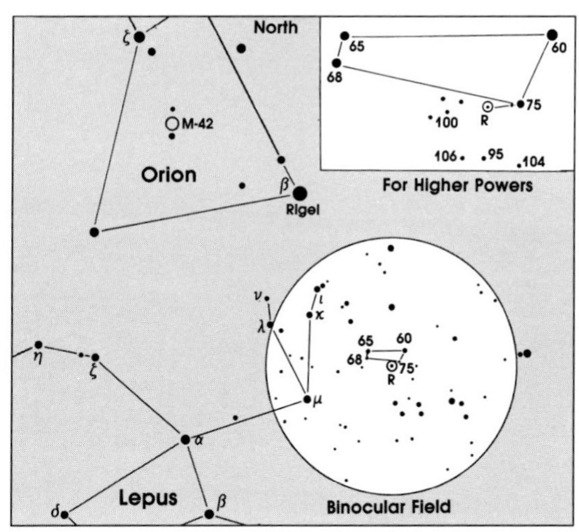

436 Days 045514 R Leporis
Range: 6.0–9.7

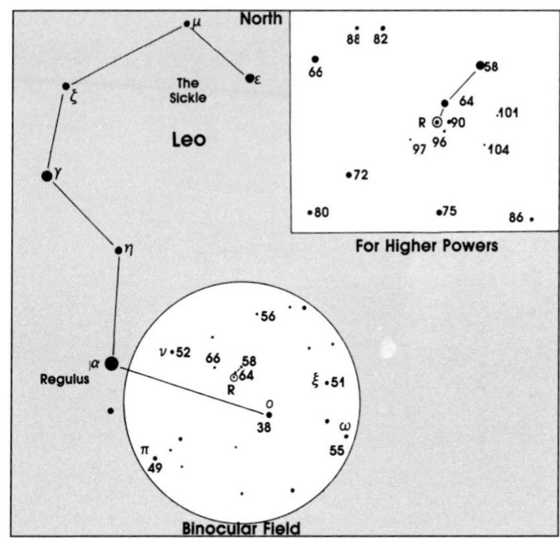

313 Days 094211 R Leonis
Range: 5.0–10.2

CHI CYGNI

For a few weeks around the time of its maximum brightness, this variable is often conspicuous in the long arm of the Northern Cross, where it may be found nearly on a line between Albireo and Gamma. It is classified as a long-period variable since its period of 46 days is one of the longest of all. Discovered in 1686, Chi is one of the "old timers." Its range of ten magnitudes is also one of the most extreme of all variables, making this star some 10,000 times brighter at maximum than at minimum, although at some maxima the star doesn't reach magnitude 6.0.

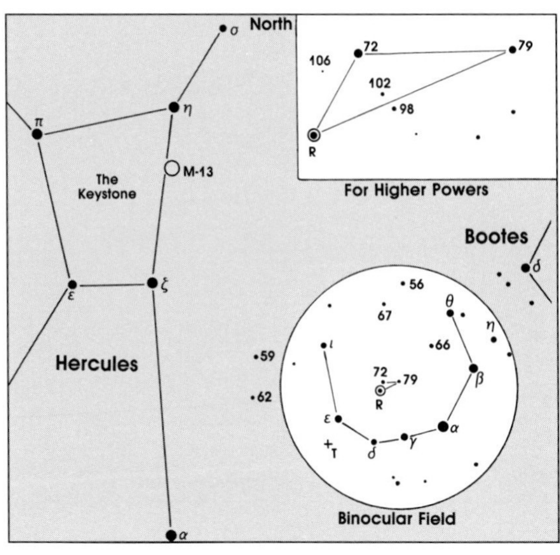

Irregular 154428 R Coronae Borealis
Range: 5.9–15.0

R LEPORIS

This remarkable "Crimson Star" was first noticed in 1845. Its deep red color is quite pronounced in binoculars. It is another long-period or Mira-type variable, and as you will note, its period of 436 days is even longer than that of Chi Cygni. You should be able to follow it through most of its cycle with 7x binoculars, though its low declination in the south may make its minimum phase rather difficult because of horizon haze.

R LEONIS

This Long Period variable can be followed through all of its brighter range with binoculars; its reddish color sets it apart from the other stars in the field. R has always been a favorite star of mine, for it was the first variable I ever found. It's also one of the easiest to find.

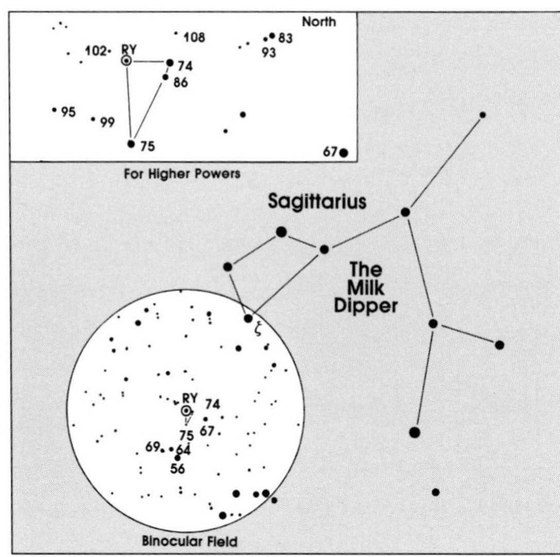

Irregular 191033 RY Sagittarii
Range: 6.1–14.0

R CORONAE BOREALIS

On the basis of performance, this is one of the most interesting stars in the sky. It seems to be totally unpredictable with no definite period. Although it usually can be found at its maximum brightness of about magnitude 6.0 and has been known to remain there for ten years at a stretch, it can also suddenly drop out of binocular vision down to magnitude 14.5 and hover there for several months before gradually rising to 6th magnitude again. The spectroscope reveals that R has an excess of carbon in its atmosphere, and it has been suggested that this may be a factor in the star's behavior.

R is found near the center of the circlet of the Northern Crown and is usually well within the power of your binoculars. Check on this star every night and keep a record of any changes.

RY SAGITTARII

This irregular variable has all the characteristics shown by R Coronae Borealis, though it doesn't seem to fade quite so low at minimum. It also shows a slight secondary variation of about half a magnitude over a period of around forty days, though it takes a keen eye to catch this. Although RY is just as interesting to watch as R Coronae Borealis, relatively few observers follow it. In our latitude the star's brief sweep across the southern sky never carries it much above the haze of the horizon.

R HYDRAE

The two lower bright stars in Corvus, the Crow, point directly to Gamma Hydrae, and R follows it by about two degrees. This is a Mira-type variable whose period has decreased by 100 days since its discovery in 1704. At maximum, this red star can be picked out by its color alone. Virtually all of its cycle is within range of binoculars.

TX DRACONIS

TX Draconis is a reddish, semiregular variable that is easily found, and its range of two magnitudes is always within reach. It is always above your horizon.

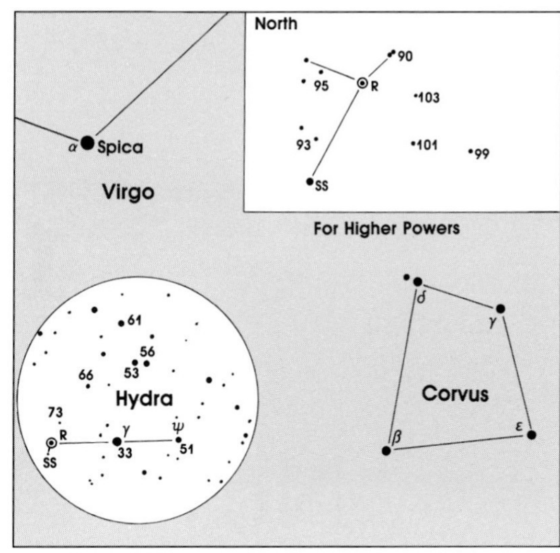

405 Days 134<u>22</u> R Hydrae
Range: 4.2–9.5

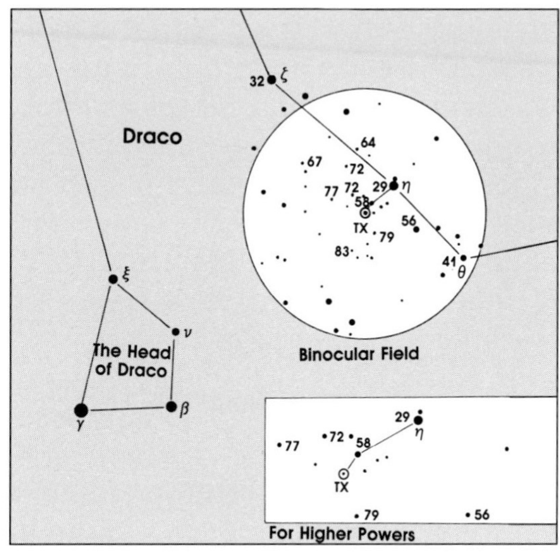

79 Days 163360 TX Draconis
Range: 6.8–8.1

11

Novae and Nova Hunting

Novae are closely related to variable stars. The terms new stars and temporary stars, which are often applied to them, suggest their ephemeral visibility. After their initial explosive outbursts, these amazing stars gradually diminish in light until, after a year or several years, they settle back to their original status as faint telescopic stars some nine to thirteen magnitudes below their maximum brightness.

The first recorded nova blazed out in Scorpius in 134 B.C. Its sudden appearance convinced the Greek astronomer Hipparchus that stars could change. Six years later, he had completed a catalog of 1,025 of the brighter stars so future observers would have a means of recognizing any variations that might take place.

In November 1572, the brightest nova on record appeared in the constellation Cassiopeia. It is usually known as Tycho's Star after its famous observer and recorder, Tycho Brahe, who thus describes his first view of the spectacle: "One evening when I was considering, as usual, the celestial vault, the aspect of which is so familiar to me, I perceived with indescribable astonishment a bright star of extraordinary magnitude near the zenith in the constellation of Cassiopeia." According to Tycho, this star could be compared to the planet Venus when it is nearest the Earth. It remained visible to the naked eye for seventeen months.

The next bright nova in the skies was the so-called Kepler's Star of October 1604, which became brighter than the planet Jupiter and remained so for over a month. The summer key map (Chapter 7) shows the location of this long-ago explosion in southern Ophiuchus. Following Kepler's Star, no novae of 1st-magnitude brightness appeared for almost

300 years. This long drought came to an end in the opening year of this century with the discovery, on February 21, 1901, of a nova not far from Algol in the constellation Perseus. In less than four days, Nova Persei rose thirteen magnitudes to equal the brightness of neighboring Capella. Today it is again a 13th-magnitude star that occasionally fluctuates by a magnitude or more.

The evening of June 8, 1918, marked the arrival of the brightest nova since Kepler's Star. That occasion has become one of the treasured memories of my own youthful nights, as I was one of a host of stargazers who witnessed the performance. This nova reached its maximum on June 9, when it equaled Sirius, the brightest of the stars.

In later years, other novae have visited our skies. These always-welcomed guest stars came in 1920, 1925, 1934, 1936, 1942, and 1975, when a nova of magnitude 1.8 suddenly blazed out about 5 degrees northeast of Deneb in the Northern Cross. When darkness fell on the night of August 31, it was quickly spotted by scores of stargazers who were favored with clear skies.

RECURRENT NOVAE

All the novae thus far described have been content to make one brief burst of glory and then retire to pass their golden years in quiet obscurity. There is, however, a small group of these star performers who have made an encore and, in some cases, several repetitions of their act. U Scorpii has thus far exploded in 1863, 1906, 1936, and 1979. RS Ophiuchi has equaled this with outbursts in 1898, 1933, 1958, and 1967. T Coronae Borealis, normally a 10th-magnitude star lying just outside the horseshoe figure of the Northern Crown, rose to 2nd magnitude in 1866; eighty years later, in 1946, it heard the call again and flared up to 3rd magnitude. These three recurrent novae are beyond the reach of your binoculars except when they are bright, but I have marked the locations of the latter two on the key maps so you may watch for any future flare-ups.

SUPERNOVAE

In the chapter on the autumn stars, I mentioned the supernova that appeared in the galaxy M31 in 1885. A supernova shines 10,000 times brighter than the average nova. Every year a number of supernovae are found by photography in distant galaxies, particularly those in the Virgo and Coma Berenices fields. In some of these faint galaxies, the newcomer, when at maximum brightness, will give out more light than the entire galaxy in which it appears. Within historic times, three supernovae have exploded right here in our Milky Way Galaxy. The first of these blazed out in Taurus in 1054 at the spot where we now find M1, the Crab Nebula, a faint binocular object among the winter stars. Tycho's Star of 1572 and Kepler's Star of 1604 were also supernovae in the Milky Way.

NOVA HUNTING

Nearly every year a nova bright enough to be seen with the naked eye is found somewhere in the sky, and many times it is the stargazing enthusiast who finds it. Here is a still uncrowded field for the persistent observer who sweeps the skies at every opportunity with both the naked eye and binoculars.

To the novice, the possibility of recognizing one more star among the host of permanent residents of the sky may appear to be quite hopeless, but in actual practice it's simple. For one thing, the nova hunter doesn't have to search the entire sky but only that portion where game is most likely to be found. Just as the bird watcher in search of waterfowl or shorebirds turns his steps toward lake or stream, so the seeker of strange stars directs his eyes or his binoculars along the mainstream of the skies—the Milky Way.

When we view the broad band of the Milky Way, we are seeing an edge-on section of our own Galaxy, and here, almost in a single plane, we can scan most of its stars. The vast majority of recorded novae have appeared in a zone about 10 degrees wide lying on either side of the galactic equator, the waistline of the Galaxy. Our search for novae, therefore, instead of covering the entire sky, narrows down to a systematic nightly inspection of the slender strip of Milky Way that lies above your horizon.

It goes without saying that to find a new star you should know the old ones. This can only be done by frequent exposure to them. In the pages devoted to the seasonal appearance of the skies, I have stressed the importance of learning the constellations not as formless groups of stars but as definite star patterns that you can quickly recognize each time you see them in the sky. Such patterns may well be of your own devising, with the principal stars of each group connected with imaginary lines to form some figure or design of your own choice.

Frequent inspection of these patterns will soon fix them in your mind. I find that as I sweep my eye along the Milky Way each of my star patterns passes in succession before me, each one so clear and definite from long association that any stranger within that design disturbs my mental image and automatically stops my eye. The culprit, be it nova, planet, or slowly creeping man-made satellite, is quickly located. Simply stated, if your star pattern is a square, a pentagon will tell you something is amiss.

Thus far, virtually all bright novae have been first sighted with the naked eye. Today, a systematic search for fainter novae is being made with binoculars in the hands of dedicated stargazers.

Nova hunting with binoculars is naturally more difficult and time-consuming than with the naked eye because of the smaller field of view and greater number of stars that binoculars reveal. Here the naked-eye constellation patterns are of little or no assistance; you must concentrate on a careful search of a small, specific area of sky. It may be a strip no wider than your field of view and no longer than your photographic memory. In this bounded bit of Milky Way you must again create your own star patterns to assist you in the prompt detection of a possible intruder.

Just what area or areas of the Milky Way you select for a nova search program is a matter of choice. Judging by past appearances, the region along the median line of the galaxy from Cygnus through Aquila, Scutum, and Sagittarius would be the most promising territory in which to stake one's claim. Nova hunting with binoculars is still in its pioneering stage in this country, and you will find little competition for likely claim sites.

12
The Sun

I can't write a book that claims to be a guide to the stars without paying tribute to the Sun—the nearest, and brightest, star and the only member of that uncounted throng that we know, with first-hand certainty, to be the sole support of a large and varied family. Among these dependents are the nine major planets (Mercury, Venus, Earth, Mars, Jupiter, Saturn, Uranus, Neptune, and Pluto) as well as a multitude of tiny bodies known as asteroids. As members of the Sun's family, all but two of the major planets—Mercury and Venus—have families of their own in the form of moons or satellites that circle about them even as they orbit the Sun. Finally, as part of this entourage are those small interlopers, comets and meteors.

This family of the Sun is not noted for togetherness, for its members are scattered all about to a distance of well over four billion miles from the central hearth fire. The only tie that binds them is gravitation. Happily, nearly every member, save for the wayward planet Pluto and the smaller asteroids, can be captured with your binoculars. I shall begin a roll call of this family with the Sun.

The Sun is a gaseous globe 865,000 miles in diameter, which is large enough to contain 1.3 million Earths. If Earth were placed at the center of the Sun, the Moon could revolve about it at its customary distance and still reach only a little more than halfway to the Sun's surface. Although the Sun is a mere pygmy compared to red giant stars such as Betelgeuse, Antares, and Mira, it is much hotter than these low-density stars. The temperature on its surface is about 6,000° C, which is cool compared to the temperature at the Sun's center, an estimated 15,000,000° C.

For the binocular observer the most interesting feature of the Sun is the parade of sunspots across the solar disk as the Sun slowly rotates on its axis. Sunspots are believed to be magnetic storms on a scale so vast that many of them are far larger than Earth. These spots appear intensely black to the eye, but this is only by contrast with the brighter surface surrounding them.

Let me strongly emphasize that you should never observe the Sun directly through any optical instrument. The danger of permanent damage to your sight can't be overstated. The safest way to observe the Sun is to mount your binoculars on a sturdy tripod, remembering to cover one of the objective lenses so no one can accidentally look through it. Next, set up a piece of white cardboard several inches behind the other eyepiece so you can project the Sun's image onto the board. Once set up, you'll be able to see sunspots easily.

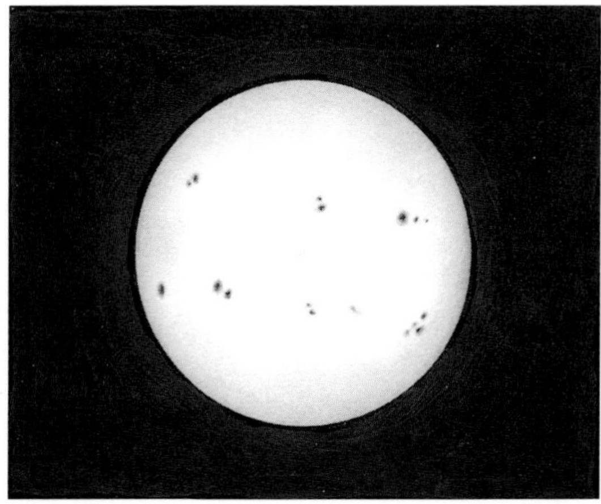

The Sun near Sunspot Maximum

You will usually find some sunspots on your first look, although not always, for sunspot activity goes through a more or less regular cycle of about eleven years. The last maximum occurred in 1991, and the next will take place around 2002. At minimum, there may be days or even weeks when no spots, large or small, are visible. But at maximum, you may even find a double row of spots, both singly and in groups, scattered all across the solar disk. Spots are partial to the Sun's middle latitudes. They are never found near the poles and only rarely appear on the equator. Their latitude is subject to the sunspot cycle. At the beginning of a new cycle, the spots appear about 35 degrees north and south of the Sun's equator and gradually appear at lower latitudes as the cycle progresses.

Stephen Wingreen

A solar eclipse is one of the sky's most fascinating events.

Because the solar axis inclines 7 degrees to the plane of the Earth's rotation, the Sun's equator stretches in a straight line across the solar disk only in the months of June and December, when the Earth crosses the Sun's equatorial plane. As we view the sunspots from Earth, they seem to move across the Sun from east to west—always parallel to the Sun's

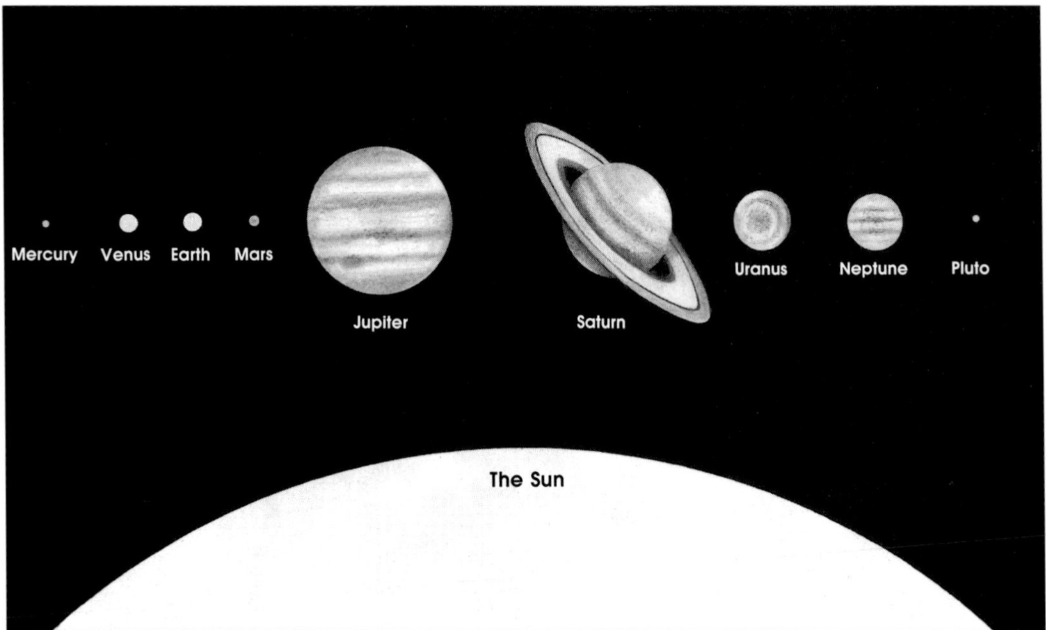

Relative Sizes of Planets and the Sun

equator. This is an apparent motion, however, for we are really watching the Sun's slow turning on its axis. By timing these apparent sunspot movements, astronomers learned that the Sun could not be a solid body, for spots in different latitudes take different times to cross the Sun's disk. A spot at the Sun's equator makes a complete revolution in 25 days, while one in mid-latitudes requires some two days longer.

Try sketching a large sunspot group as it appears through your binoculars on the cardboard. Follow it from day to day and carefully note its changes in form and size. Quite often such a group can be tracked all the way across the solar disk until it disappears at the western edge. If after a lapse of about two weeks it has survived its unseen passage across the Sun's opposite side, it will reappear at the Sun's eastern limb to begin another crossing. One exceptional group was observed for a year and a half, or some 36 crossings, though the average life of the larger groups is about two or three months. Many spots, however, will fade out and disappear even as you observe their initial crossing.

Should you wish to observe and record the life history of a sunspot group as it unfolds before you from day to day, be on the lookout for some of the following characteristics and peculiarities. A sunspot consists of a very dark nucleus called the umbra, which in a large spot may be some 40,000 miles in diameter. This is bordered by a lighter zone known as the penumbra, which can be as much as 100,000 miles across. Often associated with sunspots are the faculae, exceptionally bright areas best seen when near the slightly darker edge of the Sun's disk.

A typical sunspot group begins in two parts with their centers lying in a nearly east-west line. Of the two parts, the western (on the right) is usually nearest the Sun's equator. Such a group may gradually shrink and disappear or it may grow still larger, in which case the eastern part may divide into smaller spots and slowly disappear while the western part becomes a large round spot that may persist for many crossings of the Sun.

In 1843, Heinrich Schwabe, an amateur stargazer of Dessau, Germany, after twenty-seven years of watching and recording sunspots, announced that the spottedness of the Sun follows a periodic cycle that has been determined to be about eleven years. It is near the times of maximum activity—when a large sunspot group crosses the Sun's meridian and is turned directly toward us—that Earth is most likely to have magnetic storms and active displays of aurorae, "northern lights," as well as unusually strong radio interference.

I heartily recommend sunspot watching as a challenge for binocular observers. In the broad sense, this is stargazing just as truly as sweeping through Orion or the Milky Way.

The American Association of Variable Star Observers (address given in Chapter 10) sponsors a Solar Division for those interested in observing and recording sunspots.

13
The Planets

The previous chapter on the Sun listed the names of the major planets in the order of their distance from the Sun. With the aid of binoculars, we will get a closer look at each of them except small Pluto, which is just too little and too far away. An acquaintance with our brother and sister planets is more than just a matter of family pride and kinship, it is a worthwhile accomplishment that a stargazer will be able to use to recognize the bright wanderers that so frequently disturb the familiar figures of the constellations.

It is, I admit, a bit more difficult to get to know the planets than their "lookalikes" among the stars. This is because the stars are fixed in place; they never leave the star pattern that surrounds them. Antares, for example, will always be the heart of the Scorpion, whose long, curved tail dangles down to the horizon, but the planet Mars—often Antares' identical twin in color and brightness—may appear in any of a dozen constellations. Aldebaran never leaves its familiar place in the V-shaped Hyades, but the planet Saturn, which equals it in brightness when its rings are turned edgewise toward us, has no permanent star pattern to point it out. The best guide is a thorough knowledge of the brighter stars that lie along the interconstellation route traversed by the planets.

This planetary pathway is called the zodiac, a narrow strip of starry background that extends all the way around the sky and against which the Sun, Moon, and planets appear to move as we view them from Earth. This band of sky is 18 degrees wide with the ecliptic, the apparent path of the Sun, as its center line. Only the Sun and Earth, as viewed from each other, stay strictly on this center line, for the ecliptic is the plane of the great circle that Earth draws around the Sun each year. Each of the principal planets, as well as our own Moon, travels in a plane tilted more or less to the ecliptic but always within the borders of the zodiac.

MERCURY

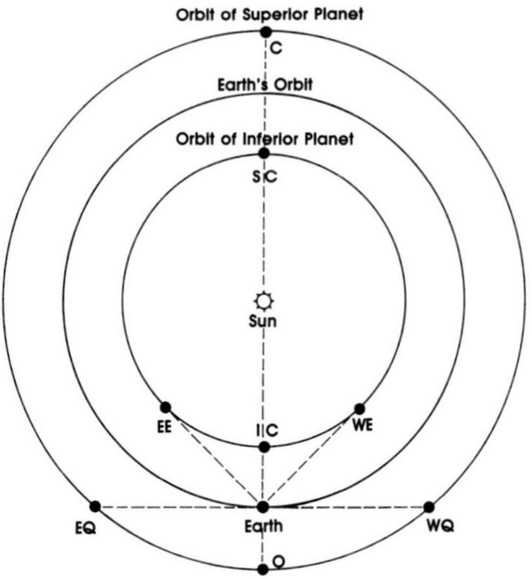

This drawing is a bird's-eye view of three different planetary orbits. The inner circle represents the orbit of an inferior planet, such as Mercury or Venus, that revolves inside the Earth's orbit. The outer circle is the orbit of a superior planet, one that revolves outside our orbit.

An inferior planet cannot be seen from Earth at either IC or SC. We view these planets best between IC and EE (after sunset in the western sky) or between IC and WE (before sunrise in the eastern sky).

A superior planet cannot be seen at C. Such a planet is nearest the Earth and therefore brightest at O when it is opposite the Sun in the sky, though it can be well observed from EQ to WQ, when it is at right angles to the Sun.

In any planetary book of records, the planet Mercury would be a frequent entry. It is the nearest planet to the Sun—some 36 million miles distant. Along with Venus it is one of the hottest planets, with a temperature of some 450° C on the sunward side, but the nighttime side may well be the coldest of all at -180° C. With a diameter of 3,031 miles, it's the smallest of the major planets, and its orbital speed about the Sun of 30 miles per second makes it easily the swiftest. Radar observations indicate that Mercury's day equals 58.7 of our days and, until Pluto came along and challenged it, Mercury also claimed the most eccentric orbit and the orbital plane most inclined to the ecliptic.

Mercury is the most elusive of the brighter planets; many who are otherwise quite familiar with the sky have never seen it. This is because in our temperate latitudes it can never be seen against a completely dark sky. Even at the times of its greatest elongations east or west it is never more than 28 degrees distant from the Sun, which is less than three wide-fields in binoculars. Despite these restrictions, finding Mercury can be simple when the time and conditions are favorable. I have often seen it with the naked eye glowing like a distant beacon in a clear western twilight sky less than an hour after sunset in early spring. Mercury can be conspicuous, for at its brightest it appears to the naked eye almost as bright as Sirius, which at times is also shining not far distant. Most often, however, Mercury shines at about the brightness of Capella or Arcturus.

Apparitions of Mercury are particularly good when they occur in the western twilight sky in March and April and in the morning sky before sunrise in September and October. Note the point on your horizon where the Sun sets or rises; this will be one point on the ecliptic. At the time of an elongation, you should be able to easily find Mercury with binoculars not far east or west of this point.

In a number of ways Mercury is quite similar to our Moon. Though it is somewhat greater in size and density than the Moon, it, too, is a poor reflector of the sunlight that falls on it. Space probes have shown that its surface is a pitted, crater-pocked terrain much like the lunar landscape. Mercury also has phases that are, in miniature, just like those of the Moon, though they can't be observed with low-power binoculars. Finally, Mercury sometimes comes directly between Earth and the bright face of the Sun, as does the Moon.

These infrequent transits across the Sun provide the greatest thrill for the binocular observer of the planet. To prevent eye injury, the same precautions must be taken here as in observing sunspots. When in transit, the little planet looks much like a perfectly round sunspot without the usual penumbra. Like a sunspot it moves across the Sun from east to west since transits occur only when the planet is in retrograde, or westerly, motion. About thirteen transits of Mercury occur each century, and these can only take place in May and November. The last transit was on November 6, 1993, and the next will be on November 15, 1999. Each transit lasts about five hours. I wish you cloudless skies for these occasions.

VENUS

Venus, the second planet from the Sun, is only slightly smaller in size, mass, and density than Earth. It is also our nearest planetary neighbor. The planet lies about 67 million miles from the Sun. Like Mercury, Venus shows phases as it travels around the Sun, and in this case they can be seen in binoculars. The crescent phase is most easily seen when the planet is about 35 days before or after inferior conjunction, when it passes between Earth and Sun. If you follow the course of the planet with your binoculars, you'll notice a great variation in the size of the disk as it travels around the Sun. This is because Venus is six times farther away from us at superior conjunction than it is at inferior conjunction, when it comes closest to Earth.

At its brightest Venus is at magnitude -4.5; next to the Sun and Moon it's the brightest object in the sky—a fact that for years has made it a prime target for those who have a penchant for seeing flying saucers. Because of its great brightness, you will probably want to observe Venus in daylight or twilight rather than at night. It isn't difficult to locate this planet in afternoon daylight if you mark its course the day before during evening twilight with fixed terrestrial objects, such as trees and buildings, and then extend this course 15 degrees eastward for every hour before twilight begins. A still simpler method is to get the planet in your binoculars in the eastern sky before dawn when the sky is still dark. With your binoculars mounted on a tripod you can follow the planet long after the Sun is high up in the sky.

Despite its closeness to us, no one has seen the surface of Venus from Earth—even with the most powerful telescopes—for it is always covered with a heavy layer of clouds. A Venusian day lasts 243 of our days, which is longer than a Venusian year of 225 days. Curiously, Venus is the only planet that rotates clockwise on its axis.

Space probes have measured the surface temperature of Venus as 750° C; its atmosphere is mostly carbon dioxide. It is definitely no place for stargazers, as the thick cloud cover would prevent them from seeing any stars or even their near neighbor, Earth, the temperature would melt lead, and the surface pressure would crush steel.

Venus, like Mercury, occasionally transits the Sun, and at such times its fairly large black spot can be easily seen with the naked eye. You'll have plenty of time to get ready for the next transit of Venus, which won't occur until June 8, 2004. (The next transit after that will be on June 6, 2012, and then there won't be another one for over a hundred years.) It is doubtful if any astronomer living today has ever witnessed a transit of Venus, as the last one took place in 1882.

The first authentically observed transit was one of Venus, which occurred in 1639. It was predicted by Jeremiah Horrocks, a young English clergyman, and as it occurred on a Sunday, his ministerial duties prevented him from seeing the beginning of the transit; but his friend Crabtree, whom he had informed of the impending phenomenon, witnessed the entrance of the planet on the solar disk, and both young observers saw it before the end.

I like to picture in my mind the tense drama taking place in the little Lancashire village of Toole on that Sunday afternoon as young Jeremiah, then but twenty years old and serving his first pastorate, hurried through the discharge of some duty in connection with his ministry, while all the time his thoughts kept straying to the clarity of the sky. From time to time he would look up anxiously as drifting clouds dimmed the sunlight streaming through the stained glass windows of the church. Those same clouds could prevent his witnessing a sight that no eyes in all the world had ever yet seen.

Jeremiah's mind mind was also on the flying minutes and this added to his worries, for according to his calculations at three o'clock that afternoon the small black image of the planet would start to creep across the Sun. Right now, he told himself, this transit may already be in progress.

When he finally arrived and got the Sun in focus, Jeremiah found the planet already on its westward trek across the bright face of the Sun. He was able to watch the transit for half an hour until the Sun set on that brief but momentous December day. Needless to say, young Horrocks was overjoyed that his calculations had proved to be completely accurate and that he had witnessed that rarity of rarities—a transit of Venus. He knew full well that he would never see another, for the next would not occur for more than 120 years.

On that Sunday afternoon in 1639, two centuries before Darwin and *The Origin of Species,* we have an early example of the conflict between religion and science. And it was solved quite happily—by compromise.

MARS

Mars, the fourth planet from the Sun, has a diameter of some 4,217 miles, a little more than half that of Earth. In spite of its small size, it becomes, under favorable circumstances, the brightest object in our night skies after the Moon and Venus. In binoculars, the orangy-red color of the planet is pronounced, though with low power it is difficult to make out any definite disk.

Mars revolves about the Sun in a path outside the orbit of Earth, and therefore we can never see it in a crescent phase as we do Mercury and Venus. Should you ever find Mars comparable to Jupiter in brightness, it means that our neighbor planet is near opposition, the point where the Earth is almost between Mars and the Sun. It is at opposition that Mars comes closest to the Earth in its 687-day trip around the Sun. However, because the Sun is not at the center of Mars's orbit, the red planet comes nearer the Earth at a perihelic (nearest the Sun) opposition than at an aphelic (farthest from the Sun) opposition.

The distance between Earth and Mars can vary from about 36 million miles at the very best oppositions to about 63 million miles at the least favorable. On the night of August 11, 1971, Mars was at its nearest and was brighter than the planet Jupiter. Such a close approach won't occur again until the year 2003.

At any opposition it is interesting to follow closely Mars' nightly trip against its starry background. As the time of opposition approaches you'll notice that the normal eastward movement of Mars among the stars is slowing down. Eventually it will seem to come to a complete stop, then reverse its direction and move backwards, or toward the west. After a few weeks this backward, or retrograde, motion slows to a halt and Mars turns eastward, resuming its normal motion. The explanation of this erratic course in the sky is a simple one. They occur at a time when Earth and Mars are nearest. Earth, being nearer the Sun, moves more rapidly in its orbit than Mars and, at opposition, is passing Mars, thus making the red planet seem to be moving backward against its starry background—just as a speeding car, in passing a slower one on the highway, makes the latter seem to be moving in reverse against the landscape in the distance. If you plot the motion of Mars during its retrograde phase you will find that due to the inclination of its orbit, it makes a slender, well-defined loop as seen against the stars.

This demonstrates the great advantage of binocular over naked-eye observing. The binoculars reveal a background liberally sprinkled with stars, among which the planet clearly shows every detail of its changing path, its slowing motion, its acceleration, and the brief stationary pause at either end of the narrow retrograde loop.

Mars rotates on its axis in 1.026 days. The planet's atmospheric transparency has made Mars the most critically observed planet. Nearly a century ago, a number of observers, both in America and Europe, claimed to see a network of dark lines geometrically placed all about the circumference of the planet. These narrow lines, both curved and straight, came to be called "canals" and were assumed by some to be artificial waterways constructed by a race of intelligent beings in an effort to conserve and distribute the water from the seasonal melting of the snow or ice in their polar regions.

Among those who claimed to clearly see these markings was a wealthy amateur astronomer, Percival Lowell of Boston. The brother of poet Amy Lowell and A. Lawrence Lowell, president of Harvard University, he built and equipped a private observatory in the clear skies of Flagstaff, Arizona. From this mile-high station soon came drawings showing hundreds of vegetation-bordered canals and oases in great detail.

Around the turn of the century, Lowell published three books that dramatically described Mars as the abode of an advanced civilization. Few astronomers of the time agreed with these purported sightings. Professor E. E. Barnard, who was said to have had the most acute eyesight on record, was never able to see a single canal. But the public loved the Lowell books and the case for the canal-digging Martians got a terrific boost.

Inspired perhaps by Lowell's claims of a nearby neighbor culture in our solar system, another Mars-related book appeared in this same era that had an even greater impact on both the reading and listening public. In 1898, in the midst of the canal controversy, English novelist H. G. Wells published his early science-fiction thriller, *War of the Worlds,* in which an armada of space ships from Mars arrives on Earth and begins to take over our planet. We had no defense against the sophisticated weapons of the invaders, but neither had they any resistance to our germs, and so they perished to the last Martian.

Mars and Its Two Moons Phobos (left) and Deimos (right)

War of the Worlds was Earth's first science-fiction involvement in interplanetary warfare. It also proved to be the most publicized when, forty years later on the night of October 30, 1938, the story was dramatized on radio by Orson Welles. Fortunately, I heard the program from the beginning and knew it was fictional. Some uncritical or uninformed listeners had a night of sheer terror—especially those living in New Jersey, where the supposed invaders landed.

As it turned out, spacecraft from Earth vanquished the Martian canals. Beginning in 1965, three Mariner space probes made close flyby approaches to Mars, and in 1971 Mariner 9 entered an orbit around the planet. The many photographs taken by these unmanned probes failed to show a surface crisscrossed with interconnecting canals as Lowell would have had it, but did reveal one thickly strewn with craters much like those that cover the surface of the Moon.

As a final blow to the Martian cause, the Mariners also found an extremely low temperature, no surface water, and an atmosphere composed of 95 percent carbon dioxide. Two later Viking vehicles soft-landed on the surface but found no organic matter in the soil.

In 1877 two tiny moons were discovered circling closely about Mars. Phobos, the inner one, is only 5,700 miles above the planet's surface and crosses its skies three times a day from west to east. Deimos, the outer moon, is some three times more distant from the planet. Both of these moonlets appear on Mariner photographs as heavily cratered masses of rock or metal that are roughly potato shaped rather than spherical. Phobos is about 16 miles long and Deimos is only half as large. Both are captured asteroids.

THE ASTEROIDS

If you were to make a scale model of the solar system with the nine planets properly spaced according to their distances from the Sun, you would find a large gap between the planets Mars and Jupiter. According to astronomical logic, there should be a planet somewhere in this void. Its apparent absence so worried a number of observers in the late 18th century that they made a concerted effort to find the missing body.

On January 1, 1801, the Sicilian astronomer Giuseppi Piazzi, while engaged in mapping stars, found what proved to be at least a part of the unknown planet. It was not really much of a planet, being only 635 miles in diameter, but it was at the proper distance from the Sun and it pursued the proper path. It was given the name Ceres from the mythological goddess of agriculture.

Ceres remains the largest of the asteroids. It was followed by Pallas (diameter 335 miles) discovered in 1802, Juno (155 miles) in 1804, and Vesta (340 miles) in 1807, after which no more were found until 1845. Today asteroids are still being found on photographic plates, and well over 5,000 are now named. Most of them are quite small, ranging in size from small mountains to irregularly shaped rocks less than a mile in diameter. In many cases their orbits cross that of Earth. For example, on October 28, 1937, tiny Hermes, only a mile in diameter, came within 485,000 miles of the Earth; it could, on some future pass, come even closer to us than the Moon. Several asteroids in the late 1980s and early 1990s passed even closer to Earth.

Of the brighter asteroids, Ceres, Pallas, and Vesta are easily seen with binoculars. I have seen Vesta, the brightest of them all, with my naked eye without difficulty. Astronomy magazine often includes charts showing their locations in the sky.

JUPITER

This giant of the planets is larger than all the other planets combined. Its mass is 300 times that of Earth, and its mighty globe could contain 1,300 bodies the size of Earth. It is the fifth major planet outward from the Sun and is more than three times farther from it than its neighbor, Mars. Jupiter moves along its orbit at the leisurely speed of only 8 miles per second, which results in a year that is the equivalent of twelve of ours and conveniently places the planet in a different zodiacal constellation every year.

In sharp contrast to its long year, Jupiter has the shortest day of any of the planets, for it rotates on its axis in 9 hours 55 minutes. This is indeed surprising when you consider the immense size of the globe. Any spot on the equator of Jupiter travels at a speed of some 28,000 miles per hour. This generates sufficient force to make the giant planet a globe with a bulging equatorial belt and noticeably flattened poles.

Unlike Mars, we never see the surface of Jupiter; we see only the top layers of the clouds of hydrogen compounds that cross the planet in narrow belts parallel to the equator. By careful observations of these belts with telescopes and spacecraft astronomers have learned that the equatorial belts make a complete rotation about five minutes faster than the belts nearer the polar regions of the planet.

For an observer with binoculars, Jupiter is probably the easiest of all the planets to distinguish, for it carries its identification marks in the form of

Mansfield Clinnick

Jupiter and its four brightest moons are visible in binoculars.

four moons that circle about it in an ever changing retinue of shadows, transits, and eclipses. These four moons, discovered in 1610 by Galileo with his tiny, brand-new optic tube, are all within the powers of your steadily held binoculars, and you can find delight and fascination in following their activities from night to night as they play slow-motion hide and seek about the bright globe of the planet.

In keeping with Jupiter's size, these four moons are also sizable affairs, although your binoculars will show them only as tiny moving stars. In the order of their distance from the planet, they are: Io, Europa, Ganymede, and Callisto. Io, though a little larger than our Moon, is the most difficult to see, as it is the closest to the glare of the planet. Europa, the smallest of the four, is slightly smaller than our Moon. Ganymede, the brightest of them all, is somewhat larger than the planet Mercury; and Callisto, a rather poor light reflector, is a little smaller.

The orbits of these four bright satellites lie almost in the plane of Jupiter's equator, which also is virtually in the plane of the ecliptic, so we see the satellites nearly in a straight line drawn through the equator of the planet. While watching the transits, eclipses, and occultations of these four moons as they circle about Jupiter, you may realize that you are watching an animated miniature enactment of what you might be seeing if you stood on Jupiter watching the slow circling of the four inner planets—Mercury, Venus, Earth, and Mars—about the Sun.

Occasional reports claim seeing Jupiter's moons with the naked eye. I find it easy to believe such claims, since their magnitudes range from 5.0 to 6.0. Such sightings would present no problem if the moons were located farther from the glare of Jupiter. It would make a most interesting project to see if you could spot any or all of these satellites with your naked eye. Make a drawing of their locations in relation to the planet, then check with your binoculars to learn if you really did see them. This is a difficult test of eyesight, so to give yourself every advantage choose a time when Jupiter is near opposition in one of the northern zodiacal constellations, such as Taurus, Gemini, or Cancer, and at an hour when the planet is near the meridian. It will then be at or near its highest point in your sky and can be seen through the least possible atmosphere.

The Jovian satellite system is truly an enormous one, for the planet actually has 16 or 17 named satellites, though only the four bright moons are within binocular range. The others are all small, not over 100 miles in diameter, and can be seen only with powerful instruments.

In addition to providing uncounted hours of fascinating entertainment for generations of sky observers, Jupiter and the four bright moons made a definite contribution to exact physical science when, in 1675, the Danish astronomer Ole Roemer noted a discrepancy between the predicted and observed times that satellites were eclipsed in Jupiter's shadow. He found that these times varied with the distance between Earth and Jupiter: when the Earth was closest to Jupiter eclipses occurred 16 minutes earlier than eclipses that took place when we were on the opposite side of our orbit, some 186 million miles more distant. Those 16 minutes were the interval that light requires to cross the orbit of the Earth, making the speed of light a little more than 186,000 miles per second.

You will find the configurations and the nightly phenomena of these four bright satellites of Jupiter conveniently plotted for you in astronomical periodicals and in a number of observer's handbooks.

SATURN

Saturn is the second largest of the planets and the sixth in order of increasing distance from the Sun. When seen for the first time in a telescope of fair size and performance, it never fails to evoke an exclamation of amazement and delight from even the most blasé, for when well posed and properly positioned, it is easily the most pleasing picture in the solar system's gallery of skyscapes. Unfortunately, when seen in low-power binoculars, it is apt to be a disappointment.

I am sure that Galileo knew a measure of this same chagrin when he turned his tiny telescope to this outpost planet on those nights in 1610, for he failed to figure out the true nature of the strange three-part shape

Jim Rouse

The orange glow of Saturn appears in binoculars; large binocs are needed to hint at the rings.

he saw. When it later changed into a single ball, he gave up in despair and turned his talents elsewhere. In probing this same planet today, we shouldn't expect too much either from our eyes or our binoculars.

Like Jupiter, Saturn has a rapid rotation on its axis with a day that is only about 50 minutes longer than that of Jupiter. As a result, it has the characteristic flattened poles and bulging equatorial girth. It also is well supplied with satellites—18 named moons in all. Of these, your binoculars may give you just a glimpse of 8th-magnitude Titan, the largest, on a winter night of perfect clearness when Saturn is high up in the sky. Titan is somewhat larger than Mercury and may be the largest of all moons.

Saturn requires 29.5 years to travel completely around the Sun at an orbital speed of six miles per second. This leisurely pace keeps it wandering through the same zodiacal constellation for three or four years at a stretch. The density of the planet is less than that of water, so it would easily float in an ocean large enough to contain it.

Saturn's polar axis is inclined to the plane of its orbit even more than our axis inclination of 23½ degrees; therefore, Saturn also has seasons, which is most fortunate for Saturn-gazers on Earth: It is because of this tilt in its axis that we're treated to three-dimensional views of Saturn's most famous feature, its unique ring system. It consists of what appears to be three concentric, flattened rings fitted together, one within the other, completely encircling the planet some 9,000 miles directly above the equator of Saturn's 74,900-mile-diameter globe. The Voyager craft revealed that these rings are actually composed of hundreds of rings.

This system of rings is 170,000 miles across but less than 10 miles thick. If the length of this printed page were to represent the diameter of Saturn's ring system, the thickness of this paper is four times too great to be in scale with the extreme thinness of the rings!

During half of Saturn's long trip around the Sun, or about fifteen years, the north side of the planet and the rings are tilted toward the Earth. As the season advances, more of the ball comes into view. The rings appear to flatten out until, at midpoint, we see the planet from pole to pole and the ring, for a brief time, completely disappears. This occurred in 1995. After this, the system tilts away from us and the opposite or underside of the rings and the other pole comes into view for the remainder of Saturn's year.

My last ringless view of Saturn came in 1966. I watched it through a 12-inch refractor, and according to my records, the rings had completely disappeared when I looked on the night of October 29, though I still could see the black shadow they cast across the equatorial belt of the planet. When clear skies came again on November 1, the rings had reappeared, though not yet as a solid line but as a series of bright dashes broken by dark gaps where the ring divisions lay.

When opened to their greatest extent as seen from Earth, the rings reflect more than twice the light as the ball of the planet; therefore, Saturn's brightness as we see it is governed largely by its ring position. At its brightest, the planet may equal Capella or Arcturus, but without the ring, when the edge is turned toward us, Saturn will be no brighter than Aldebaran or Altair. The rings are neither solid nor fluid but are composed of small particles such as rocks or other aggregate revolving independently about the planet.

Among my acquaintances are bird watchers who have traveled a thousand miles to see a Kirtland's Warbler in its native haunts. The night sky has its own rare specimens that are well worth a visit to an observatory, or perhaps an evening with a friend who has a telescope. No one should pass through life without a look at Saturn and its rings.

URANUS

All the planets described thus far have been known since ancient times, for they are easily visible to the naked eye, but far outside the orbit of Saturn slowly revolve three major planets—Uranus, Neptune, and Pluto—that came to us only through telescopes.

The first of these, Uranus, was discovered quite by accident by an amateur stargazer and telescope maker who was best known among his acquaintances as an excellent oboe player in a professional orchestra.

As countless amateurs have done since, William Herschel built a telescope. With this 6-inch instrument he began to poke around the sky, and perhaps unlike the great majority of his followers, he kept a careful record of his sightings. For example, in his journal for the night of March 13, 1781, his entry was a strange note, even for an oboe player: "In the quartile near Zeta Tauri, the lowest of two is a curious either nebulous star or perhaps a comet."

For nearly a year the object was observed as a comet, but finally it was shown to be traveling in a nearly circular orbit that could only be that of a planet. Thus Uranus, the god of Heaven, was welcomed into the official family of the Sun, and William Herschel was duly knighted by King George III as the discoverer of a New World.

But just as the "New World" of Columbus had been sighted many times before by others, so too had Uranus been seen and recorded at least a score of times by four different observers before Herschel came upon it. John Flamsteed, England's first Astronomer Royal, saw it on six occasions from 1690 to 1715, and Pierre Le Monnier, a French astronomer, observed it on twelve nights between 1750 and 1771. Each time it had been recorded as an ordinary star. Uranus is the fourth largest planet and the seventh in order of distance from the Sun. Its axis is the most highly inclined of any planet, about 98 degrees; thus, when the polar axis points toward the Earth, it gives the planet the appearance of rolling along its orbit rather than spinning on its axis like a top.

Until recently, Uranus was thought to have only five moons. Herschel had found the brightest two on a single eventful night in 1787. But the count tripled over a span of three weeks in January 1986 as Voyager 2 flew by the planet and discovered ten new moons, bringing the total to fifteen. These new moons are all tiny compared with the five previously known moons—Titania, Oberon, Umbriel, Ariel, and Miranda. The smallest of the old moons is three times larger than the largest of the newly discovered satellites.

Voyager 2 had been exploring the solar system since 1977, sending back invaluable pictures and data from Saturn, Jupiter, and Uranus. The photos of Uranus showed a uniformly colored sphere with only the slightest hints of cloud features within the blue-green atmosphere. Voyager also photographed the nine thin rings circling the planet and revealed a tenth ring. The length of a Uranian day was also accurately determined from data returned by Voyager to be approximately 17 hours.

All along, Uranus has been reluctant to reveal its secrets. Not until February 1948 was its fifth moon found, and as recently as March 10, 1977, astronomers discovered that Uranus, much like the planet Saturn, has rings. William Herschel would have readily accepted the announcement of these rings, for at one time he thought he had seen such things around his planet. He never would have believed, however, the way in which they were discovered: through a 36-inch telescope in an airborne observatory eight miles above the Indian Ocean.

Like the planet itself, the rings turned up by accident. Their discovery was a by-product of an effort made by NASA's Kuiper Airborne Observatory to record photoelectrically the occultation by Uranus of an 8.9-magnitude star, SAO 158687. Forty minutes before the star was due to be occulted by the body of the planet its light suddenly disappeared for seven seconds and then reappeared. Several minutes later the star's light again blacked out and again reappeared. At shorter intervals, this fluctuation was repeated until a total of five minima were recorded.

After the fifth dip in light came a period of 30 minutes in which the star's light continued steady until it was occulted by the planet. When the star reappeared 25 minutes later there followed another period of 30 minutes of steady light followed by another series of five disappearances, but this time their intervals were spaced in exactly the reverse order. It could only mean that the shadows cast by five concentric rings had passed in succession across the open telescope. Further observations have confirmed these rings and revealed four more.

When seen at its best, at opposition, Uranus is visible to the naked eye at magnitude 5.7 and, of course, it can be found without difficulty with binoculars if you know its approximate location in the sky.

NEPTUNE

Scarcely had Uranus firmly established its status as a new world when astronomers found it straying somewhat in its path around the Sun. It was seldom precisely where their computations showed it should be. For years it moved faster in this orbit than it seemed it should; then, as though weary, it began to lag behind. Only some sixty years after discovery, the planet was off its predicted position by an apparent distance equal to about one-fifth the separation between Mizar and Alcor in the Dipper.

This was something the astronomers couldn't accept without a struggle. They reasoned that there must be an unknown body out beyond the orbit of Uranus that was preventing it from moving freely in its course about the Sun.

Of those who applied themselves to a study of these irregularities, two men, John Couch Adams of England and Urbain Jean Joseph Le Verrier of France, independently came up with correct solutions that told just where to find the unknown body in the sky. Adams made the earlier computations but, unfortunately, sent his predictions to an astronomer who failed to act on them promptly. In the meantime, Le Verrier completed his calculations and submitted them to Johann Gottfried Galle of Berlin with the following note:

> *Direct your telescope to a point on the ecliptic in the constellation of Aquarius, in longitude 326 degrees, and you will find within a degree of that place a new planet, looking like a star of about the ninth magnitude, and having a perceptible disc.*

A half-hour search on the night of September 23, 1846, showed the new planet, later named Neptune, within two Moon-diameters of the predicted point. Happily, credit for this triumph of mathematical astronomy has been shared equally by both Adams and Le Verrier.

Neptune is almost a twin of Uranus in terms of size, but it is a billion miles farther from the Sun and its day is about 16 hours long. It creeps along its orbit at about 3½ miles per second and takes 165 of our years to make a circuit of the Sun. Neptune has eight moons, one of them larger than our Moon; the next largest, discovered in 1949, has the longest month of any satellite, taking nearly a year to revolve around the planet. The others were discovered when Voyager 2 passed the planet in 1989.

Don't expect to see Le Verrier's "perceptible disc" when you search for Neptune with your binoculars, though the planet is readily visible there as a "star" of 8th magnitude. You will find its predicted path among the stars charted for each year in the astronomical periodicals and in some almanacs. Hopefully, as a star and planet gazer, you will welcome the challenge of such a search and will know the elation that attends the finding of this planet, which for more than eighty years was regarded as the outpost of the solar system.

PLUTO

Even with the discovery of Neptune, all the eccentricities in the erratic course of Uranus had not been explained. Accounting for these small perturbations required the presence of still another unknown body out in space, and not so long ago another one was found.

The story of the discovery of Pluto is, in many ways, the telling of a twice-told tale, for once again two men independently attacked the knotty problem of locating the unseen influential stranger. Once again, mathematics pointed the telescopes to the sky and, again, both targets were in close agreement. Beginning not long after the turn of the century, William H. Pickering in Jamaica and Percival Lowell in Flagstaff, Arizona, labored in this search for many years in an era when a computer was still a person equipped only with a pencil.

Not until 1930, when both men were dead, did success arrive. A comparison of two photographs taken of the same region on different nights showed a slight displacement in one faint starlike image. Clyde Tombaugh, an assistant at Lowell Observatory, photographically detected Pluto and announced its discovery on March 13, 1930, the anniversary of the discovery of Uranus and also the birthday of Percival Lowell, at whose observatory the discovery was made. As a further tribute, the first two letters of Pluto are also the initials of Percival Lowell, and these are combined in the symbol used for the planet (♇).

Pluto is a small planet, perhaps 1,500 miles across, with a density almost like that of Earth. From its distance of 3.6 billion miles, the Sun appears as no more than a brilliant starlike point of light; therefore, the temperature of this frigid little body is about -230° C. Pluto completes a trip around the Sun in 248 of our years, traveling at a speed of only 3 miles per second. Its day lasts about 6.4 Earth days long. Its orbit is so eccentric that at perihelion, when it is nearest to the Sun, it is actually closer to the Sun than Neptune is. In fact, Pluto crossed Neptune's orbit in 1979 and will remain within it until the year 1999.

In 1978 James Christy discovered a moon circling Pluto at a distance of only 12,000 miles. Inasmuch as this moon, since named Charon, is only two or three magnitudes fainter than Pluto itself, the pair constitutes nearly a double planet. At its closest approach to Earth, Pluto is no brighter than a 13th magnitude star, so don't bother to get out your binoculars. Instead, get them ready for a visit to our Moon.

14
The Moon

With the exception of the Sun, we see the Moon more easily and more often than any other celestial object, so we need no chart to find it. For several billion years the Moon has been Earth's nearest neighbor and constant companion. It is the only other world we have ever visited in person, the only one whose surface dust holds footprints from humans.

But long before that giant step was made in 1969 we had learned many of our neighbor's secrets. In the years that followed Galileo's first sighting of those distant "seas" and mountains, the earthward-facing visage of the Moon has been measured more carefully and more precisely than have many of the features of the Earth. Quite literally, we have scratched the surface of the Moon, but we still know little of its origin or early history.

The Moon spans 2,160 miles, about one-fourth the diameter of Earth. It is so large in relation to Earth that if seen from Mars or Venus this Earth-Moon system would appear as a large planet and a small planet slowly circling about their common center of gravity as they orbit the Sun. The Moon's orbit around Earth is, like all planetary orbits, an ellipse rather than a circle. Its average distance from us is 238,000 miles, though our 7x binoculars shrink this span so that we see the Moon just as we might view it with the naked eye at a distance of 34,000 miles.

Viewing the Moon is an exercise where higher-power binoculars would help, for its surface is broad enough and bright enough that the dark and narrow field of higher powers presents no problem. However, the Moon is by far the easiest of all celestial objects to observe, and your binoculars, when steadily held, have ample power to guide you on many an evening moonwalk to hundreds of sunlit craters, seas, and mountains through all the ever-changing phases of our neighbor world.

These ever-changing phases have always been a little difficult to understand. Poets and novelists have often taken some strange liberties with the

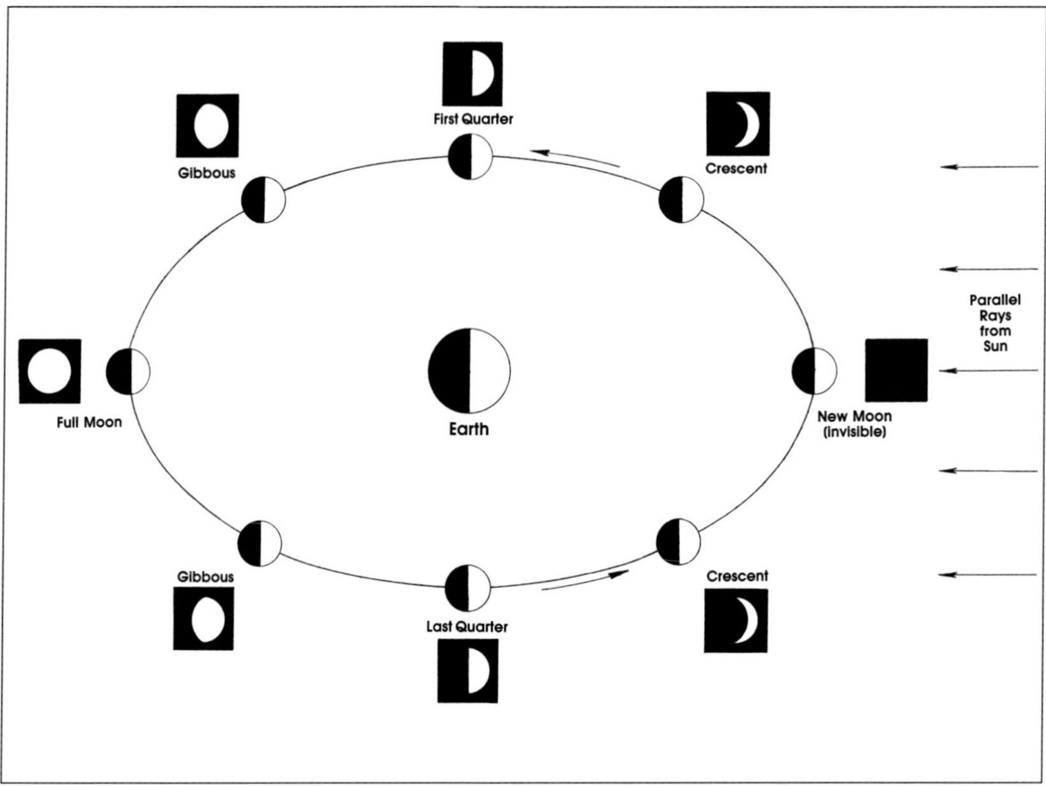

Phases of the Moon
The circling moons are shown as viewed from a point high above the orbit. The framed phases are shown as we see them fron Earth.

Queen of Night in their moonlight rhapsodies, for they have seen her slender crescent rising in the east at sunset or watched the half Moon high up in the sky at midnight. My diagram shows you just how each phase is produced, but if you, too, find it quite a problem to visualize how our Moon, a round ball, can assume so many different shapes as seen from Earth in just one lunar month, this following bit of nighttime simulation should make it all as clear as day.

First, carefully peel half an orange. This now becomes the Moon with its two contrasting hemispheres. The white half from which the peel was removed is the hemisphere that is always in full sunlight; the darker half with the peel still on is the night side of the Moon, which is always in shadow. Now push a pencil or a knitting needle directly into the orange at the line where the two halves meet. This makes a convenient handle.

Near one corner of a darkened room place a single unshaded lamp to represent the Sun. Standing near the opposite corner of the room, hold the orange by its upright pencil axis directly between the light and your head,

Paul Borchardt

The Moon offers a rich exploring ground for binoculars.

which becomes the Earth. The orange is now in the New Moon phase and quite invisible. Holding the orange rigidly before you with the peeled half facing you, slowly turn to your left. As you do so, the lighted portion of the orange will become visible to you, first as a slender crescent representing the crescent Moon as seen in the western evening sky. When you have made a quarter turn and face directly to your left, one-half of the peeled portion of the orange will now be visible, just as in the First Quarter phase. As you continue turning to your left, the white portion first assumes a football, or gibbous, phase which gets larger until, when you have made a half turn, you now have a Full Moon. Turning still farther to your left, you'll come to each of these phases in reverse order until, after one complete turn, you'll come to New Moon again.

As you performed this simple demonstration, you may have noticed some peculiar side effects that came to light in the cycle of lunar phases. The first of these occurred in the crescent phase and is known as "the old moon in the new moon's arms." You may have noticed on seeing the crescent Moon in the west just after dark that you can see the entire ball of the Moon, though only the thin crescent is brightly lighted by the Sun. This is the effect of earthshine falling on the Moon in the same way that moonlight falls upon the Earth. What you are seeing is, of course, all sunlight, but in this case it is twice reflected—from Earth to Moon and back again to Earth. In this interchange of light, the Moon comes out ahead. Its dark rocks reflect the Sun's light poorly, while the greater surface of the Earth facing toward the Moon together with the high albedo, or reflective index, of the clouds that often cover much of Earth cause the earthshine to be bright enough for you to recognize some of the lunar features on the Moon's night side with your binoculars.

As you reached Full Moon in your simulation of the lunar orbit, your hand-held Moon must have entered the shadow of your head that the lamp cast on the wall. At this point, your Moon was in eclipse, just as you may have seen the real Moon eclipsed by the Earth's shadow. However, if you want to keep this simulation true to nature, you must learn to tilt the orbit of your half-peeled Moon, for the plane in which the Moon revolves is tilted 5 degrees to the plane of the ecliptic in which Earth revolves about the Sun. Were it not for this small tilt, we on Earth would witness an eclipse of the Moon every month at the time of Full Moon as well as an eclipse of the Sun at every New Moon phase. To make this correction gradually, raise your arm as you approach Full Moon so that the orange passes above the shadow of your head. Then slowly lower your arm so the orange passes just beneath the lamp at the New Moon phase.

Obviously, an eclipse of either the Sun or the Moon can only occur when Earth, Moon, and Sun are all nearly in line with each other. These line-ups can only take place at or near the nodes of the Moon's orbit. These are the two diametrically opposite points at which the tilted orbit of the Moon crosses the ecliptic, the plane of Earth's orbit around the Sun. At these crossings, the Moon must be in either its New or Full phase for an eclipse to occur.

Eclipses, both of Sun and Moon, take on added interest when viewed with binoculars. With the Sun, projecting its image with a telescope or a pinhole camera onto a piece of cardboard will show first contact quite precisely when the black body of the Moon takes its initial nibble from the Sun. Later you can watch as the growing ball of the Moon devours any sunspots that may be present on the earth-facing solar disk. At the last moment before totality, Baily's Beads—tiny flashes of sunlight shining through the valleys and mountain peaks of the Moon's eastern edge—will appear, followed briefly by any red eruptive prominences that may be present around the Sun's rim.

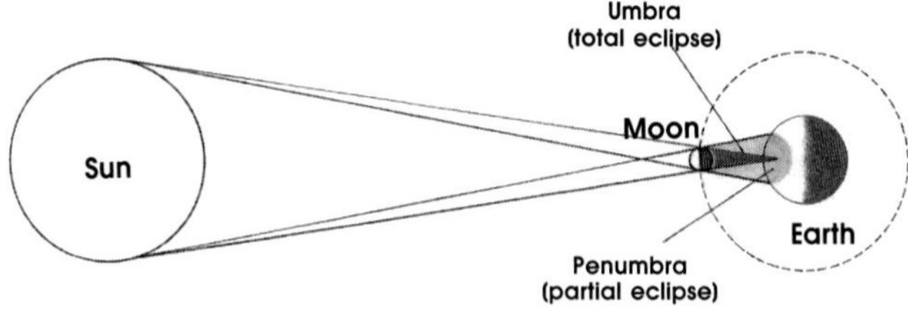

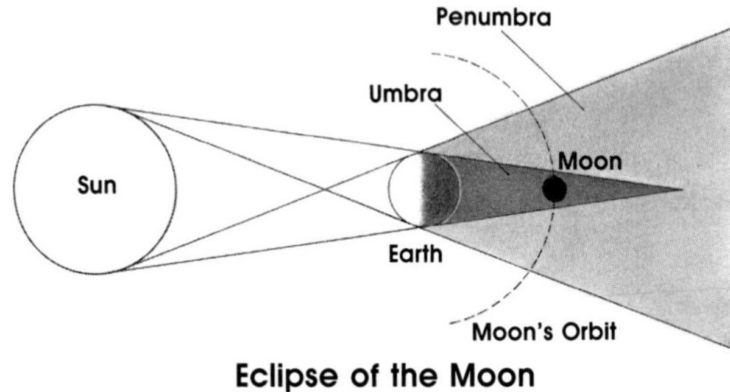

Eclipse of the Sun (Above); Eclipse of the Moon (Below)

Because the Earth's shadow contains varying amounts of light refracted by our atmosphere, the Moon usually remains faintly visible during eclipse, even when deep within the shadow of the Earth. The eclipse I watched in 1964 required some diligent searching before I was able to locate the Moon as a deep red ball with a slightly brighter rim. The Swedish astronomer Pehr Vilhelm Wargentin, for whom a crater has been named, tells of a total blackout that occurred on May 18, 1761: "The Moon's body disappeared so completely that not the slightest trace of any portion of the lunar disc could be discerned, either with the naked eye or with the telescope."

EXPLORING THE MOON

When America was a New World, it was explored from east to west. The great explorers of that era first beheld the natural wonders of this country as they approached them from the east. Our exploration of the "New World" of the Moon will follow this pattern, for we will proceed from east to west as we follow the advancing sunrise all the way across the rugged, pock-marked face of our neighbor world.

The jagged, ever-changing line that marks the division between night and day on the Moon is called the terminator. As you watch the phases of the Moon from New to Full then back to New again, you will see this slowly bending line first as a sunrise or morning terminator; then, about two weeks later, it vanishes at Full Moon. It then becomes a sunset or evening terminator that recedes nightly back to New Moon once more.

Almost any calendar or almanac will give you the date, and often even the hour, when New Moon occurs each month. The Moon at such times is, of course, invisible, except on those occasions when we are blessed with an eclipse of the Sun and can watch the Moon, a round black ball, change from Old to New right before our eyes. Some clear spring evening, shortly after sunset, try to locate a 1-day-old Moon with your binoculars. With good seeing conditions and an open western horizon this shouldn't be too difficult, for such early sightings have been made when the Moon is only eighteen hours old. The Moon has even been seen just before sunrise as an Old Moon and then again as a New Moon just after sunset on the following evening. A 2-day-old Moon is usually easy to locate in the western evening sky; you can make out a few of the lunar features as the terminator is just beginning to reveal its rugged character.

As an aid in exploring the Moon, I have drawn a series of charts showing the Moon as seen with binoculars at four stages in its cycle from New to Full. They show the more prominent features as they appear 4, 7, 10, and 14 days after New Moon.

THE 4-DAY MOON

If you have had an opportunity to view the Moon each night since the New Moon, you have doubtless already noticed that the evenings of the second and third day of its cycle showed earthshine at its best, and you may have used your binoculars to grope your way about there in the dim light from Earth. On the 4-day Moon, the most outstanding feature is Mare Crisium, the Sea of Crises. This mare (pronounced MAH-re) can easily be seen near the Moon's upper right-hand edge. It is perhaps the first lunar object to which Galileo turned his new telescope when he began his study of the Moon in 1609. In binoculars, Mare Crisium appears like an isolated inland sea, so it's little wonder that many of the early observers thought of it and other similar dark-colored areas as being bodies of water. Nor is it surprising that when old constellation maker Hevelius drew his map of the Moon he labeled each of these dark depressions a mare, a Latin word meaning sea.

In actual size, Crisium is 280 miles long by 360 miles wide—an area that would just contain the state of Missouri. Remember to look at Mare Crisium again a night or two after Full Moon. Sunset lighting will then be falling on its floor, which is the deepest of all the lunar maria (the plural of mare, pronounced MAH-ree-ah), and there you may spot a couple of small craters on this ancient bed of once molten lava. Cape Agarum, a high promontory, can be seen jutting out into the mare from its southwestern shoreline.

South of Crisium, note Langrenus. It is usually described as a walled plain because of the extent of the enclosed area. Other similar though smaller features are often known as ringed plains and craters. Inasmuch as no one knows just what caused them—either volcanic action or meteoric impact—it is largely a matter of choice which term we use.

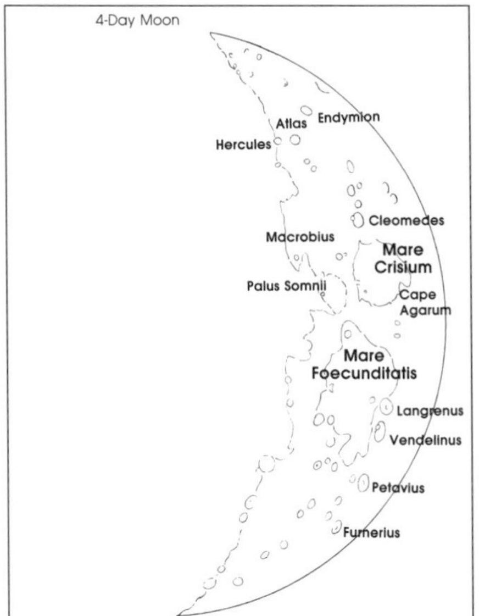

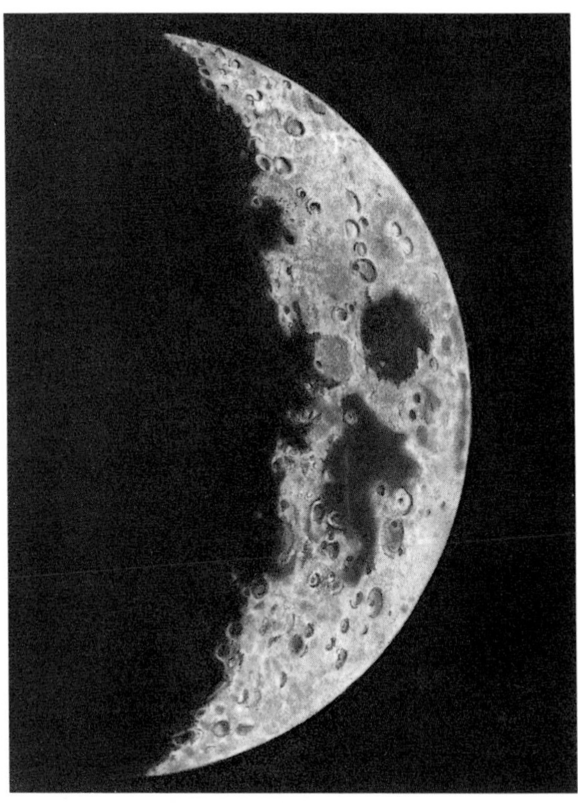

The 4-Day Moon

Still farther south of Langrenus and even larger in area is the walled plain Petavius. Note the mile-high mountain peak near the center of its floor. Just above Mare Crisium and touching its outer rampart is the oblong formation Cleomedes, whose precipitous eastern wall rises two miles above the level of the floor. Farther north and close beside the present terminator are two craters lying side by side with their outer rims just touching. They immortalize the mighty heroes Atlas and Hercules, the latter being right on the sunrise line with its floor still deep in shadow. The large dark area south of Mare Crisium is Mare Fecunditatis, the Sea of Fecundity. Note Langrenus, already mentioned, on its eastern shoreline.

THE 7-DAY MOON

The chart of the 7-day Moon on the opposite page shows our nearby satellite at its First Quarter or half-Moon phase. Both of these terms correctly describe this phase, since the Moon has now completed the first quarter of its orbit around the Earth from New to New, and at this point we can view just one half of the Moon's surface that is visible from Earth. Note that many of the ring formations of the 4-day Moon, such as

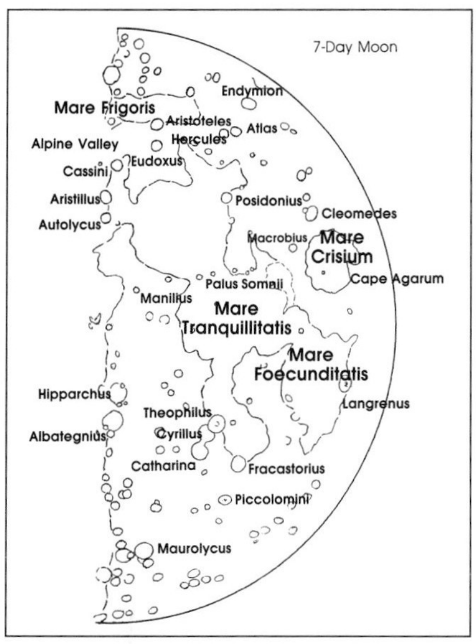

The 7-Day Moon

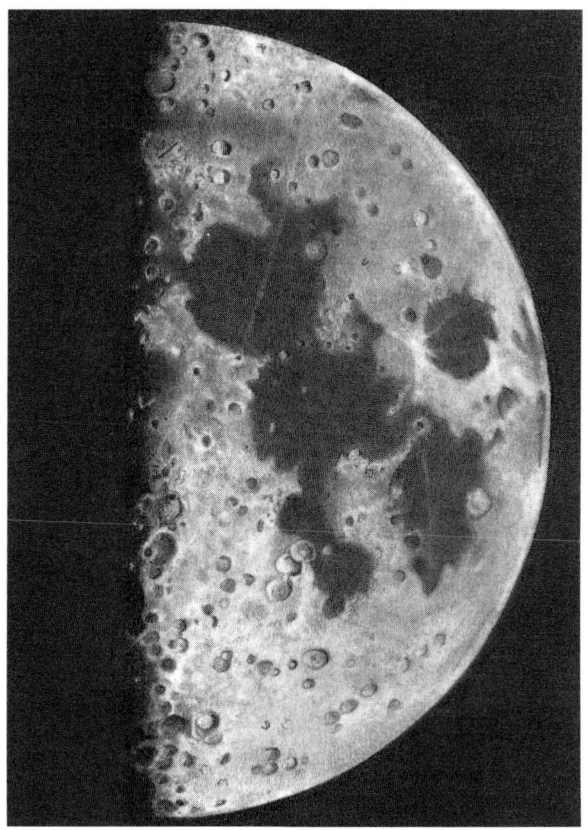

Langrenus and Cleomedes, which were then so clear and sharp, have lost much of their rugged character—along with their black outlining shadows —beneath the flatter lighting of the higher Sun.

One striking feature of the half Moon is the dark expanse of four connected maria: Mare Serenitatis, Mare Tranquillitatis, Mare Fecunditatis, and to the south Mare Nectaris, the Sea of Nectar. Look particularly at the group of three adjoining ring plains along the eastern shore of this sea of sweets. In your binoculars you will clearly see all three. Theophilus, the northernmost crater, with its prominent central peak and its high wall 64 miles in diameter, overlaps to the south the wall and plain of Cyrillus in a manner that clearly indicates that Theophilus is the more recent formation. You can also see the narrow pass or valley that connects Cyrillus and Catharina.

All along the now vertical terminator of the 7-day Moon you will find a maze of interesting craters and walled plains. Just below the center of the Moon look for another group of three walled plains that follow closely after Albategnius, the dark-floored formation that now shows as a deep crater right on the terminator. If you have time, keep watching as the advancing sunlight begins to outline the high walls of these three plains.

The 10-Day Moon

They are, from the north, Ptolemaeus, Alphonsus, and Arzachel. Do not confuse them with three similar craters we found earlier on the shore of the Sea of Nectar.

THE 10-DAY MOON

The Moon is now entering its gibbous phase as the terminator becomes a line of increasing convexity. Many of the Moon's most spectacular features are now coming into view. Your binoculars will reveal an assortment of craters, seas, and mountains that, were we closer to them, would prove as grand and awe-inspiring as anything our own Earth has to offer. Three of the Moon's most magnificent craters dominate the growing sector now lighted by the Sun. These are Plato (in the north), Copernicus (north of center), and Tycho (in the heart of a wild maze of ringed plains and craters to the south). All three are easily located with binoculars, and the fantastic terrain that surrounds each will provide many an evening of visual exploration.

Plato lies on the northeastern shoreline of Mare Imbrium, the Sea of Showers. It is a walled plain rather than a crater, for its nearly circular, dark-colored floor is some 60 miles in diameter—large enough to contain the state of Delaware. It will appear quite level in your binoculars and, unlike so many ringed formations we now have seen, it has no sign of any central peak or crater on its floor. West of Plato along this same northern shore the 10-day Moon now shows at its very best the Bay of Rainbows, Sinus Iridum. It is a broad, semicircular bay with a cape at either end that seems to guard the entrance. This bay is one of the most beautiful and earthlike formations on the entire Moon. Note how the lighted northwestern wall of the bay juts into the darkness just beyond the terminator at this phase.

At the opposite side of this 700-mile-wide Sea of Showers, the shore is bordered by the Apennines, the Moon's finest mountain range. This chain is about 600 miles long and studded with peaks that rise nearly three miles above the level of the plain. This formation can be easily seen with the naked eye.

Within the curved sweep of the Apennines and the Caucasus Mountains, which form the western border of Mare Serenitatis, note the group of three walled plains: Archimedes, Aristillus, and Autolycus. North of this trio and forming the shore of Mare Imbrium from the Caucasus Mountains to the rim of Plato, you'll find the lunar Alps. Midway in this range look for the sharp crosswise slash of the 80-mile-long Alpine Valley. This unique feature is six miles wide with a level floor that terminates among the Alpine peaks. The valley is a stretch of land, bordered by two geological faults, that has collapsed.

The rugged Apennines terminate at the crater Eratosthenes and point onward to Copernicus, the finest ring-plain on the Moon and perhaps the best known lunar feature. This grand crater, first noted by Galileo, is 54 miles in diameter and three miles deep, with inner peaks and sharply terraced walls. In a telescope the entire plain surrounding Copernicus is thickly honeycombed with hillocks and tiny craterlets.

Among the more conspicuous craters and ring-plains in the southern highland region is the ray-center Tycho, to which we will return when the Moon is Full. At the 10-day stage closely examine Clavius, an immense walled plain 140 miles across and the largest such feature on the visible lunar surface. There is a chain of smaller craters set in its broad floor.

Just before Full Moon, two contrasting craters can easily be located with your binoculars. The first of these, Grimaldi, lies near the center of the Moon's western limb. This long oval is the darkest spot on the Moon. It is believed to be a walled plain that long ago was filled with lava, now solidified. Farther north, though still in this same immense Oceanus Procellarum, or Ocean of Storms, you can easily find the crater Aristarchus, the Moon's brightest spot. So dazzling is this crater that you can see it unilluminated by sunlight. You can find it by earthshine alone when the Moon is still a slender sickle only two or three days old.

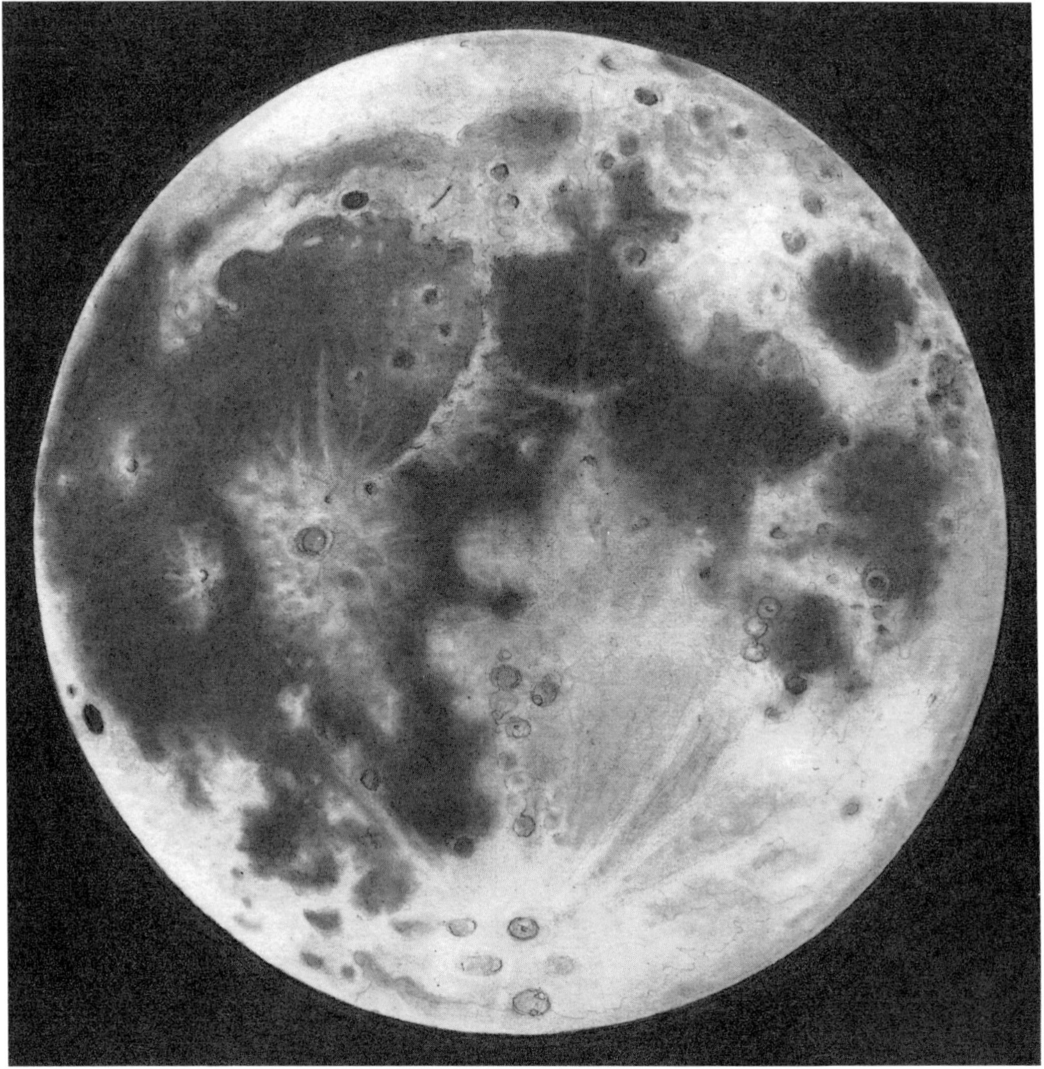

THE FULL MOON

When the Moon is Full the half that faces Earth is brightly lighted by the Sun, and no shadows can be seen. We now view the Moon with the Sun almost directly behind us. Its rays fall perpendicularly on the lunar features, and as a result many of the mountains and craters we saw so clearly at earlier phases now have almost disappeared. The terminator—that advancing line of light and shadow that has led our searching binoculars all the way across the Moon—has deserted us completely.

Because of the full-orbed lighting by the Sun, some aspects of the Moon are better observed at this stage than at any other. One of the improved features is the vast system of lunar rays that flares out in all directions,

particularly from the crater Tycho, to give the Full Moon its often noted "peeled orange" appearance. Some of these rays extend straight across the surface for a thousand miles or more. One in particular, which originates on the rim of Tycho, flares northeastward across other craters and mountains lying along its course until its gray line completely bisects the Sea of Serenity, where you can follow its straight and narrow path from shore to shore with your binoculars. Other less extensive ray systems can be seen about the craters Copernicus, Aristarchus, and Kepler.

If clear skies have permitted you to follow the ever-changing terminator of the Moon in its nightly advance from east to west, you must have often marveled at how quickly the appearance of a mountain peak, a crater floor, or the shoreline of a sea can be altered by shrinking shadows

Best Viewing Times for Lunar Features

2 days	Mare Crisium, Langrenus, Vendelinus, Petavius, Cleomedes
3 days	Cape Agarum, Furnerius, Endymion
4 days	Mare Fecunditatis, Atlas, Hercules, Macrobius
5 days	Mare Nectaris, Palus Somnii, Piccolomini, Theophilus, Cyrillus, Catharina, Posidonius
6 days	Fracastorius,Maurolycus,Posidonius
7 days	Mare Tranquillitatis, Mare Serenitatis, Aristoteles, Eudoxus, Cassini, Hipparchus, Albategnius, Manilius
8 days	Alpine Valley, Alps, Appennines, Aristillus, Plato, Archimedes, Ptolemaeus, Alphonsus, Arzachel, Walter
9 days	Maginus, Tycho, Clavius, Eratosthenes
10 days	Mare Frigoris, Mare Imbrium, Longomontanus, Mare Nubium, Sinus Iridum, Copernicus, Bullialdus
11 days	Aristarchus, Kepler
12 days	Oceanus Procellarum, Mare Humorum
13 days	Grimaldi
14 days	Lunar rays from Tycho, Kepler, Copernicus, Aristarchus, Langrenus, and Furnerius

Most of these formations are listed for the day on which they are on or near the terminator, though the mare are given for the time when the entire area of the mare can be seen.

beneath the rising Sun. For each formation on the Moon there is a certain day between the New Moon and the Full Moon phases when it is at its very best. The table above gives the Moon's age—in even days after New Moon—when many of the objects shown on my charts may be observed to best advantage.

15
Comets

Comets and their strange adventures appeal to nearly everyone. These ghostlike travelers in the night are so completely different from anything you may have encountered that I urge you to turn your binoculars on them. This may require some patience, for comets of even binocular brightness don't visit us as often as we would wish. Although two dozen or so comets are discovered every year, most of them are captured on photographic plates and few reach single-digit-magnitude brightness.

If a comet of 7th magnitude or brighter appears in some region of the sky, your binoculars will afford the best possible means of finding the invader. The wide field of low-power binoculars will give you the most comprehensive and inspiring views of the complete tail structure and the overall extent of the brighter comets that do occasionally grace our skies.

The chance that some night your binoculars may flush a comet of your own from the spangled cover of a star field is indeed a remote prospect, though it sometimes has happened in the past and, conceivably, it could occur again. Today, however, in all the dark sky regions that permit a long exposure, it is usually the photographic plate that catches the first gleam of an approaching comet.

It would seem, then, that the comet-hunter who is short on instrumental power and long on perseverance would have the best chance of making a rare find by prowling in those twilight zones where the plate, because of fogging, is quite helpless. Furthermore, if we can judge by some past instances, a diligent search—with objectives shielded by a lens hood—almost to the solar limb with the Sun just below the horizon or with its bright disk hidden behind some distant tree or building might sometimes prove productive.

Every comet for at least a portion of its long career is a captive of the Sun. The attractive power of the Sun draws a comet from the depths of space near enough to Earth for us to see it. Every comet swings about the Sun as it makes its perihelion—its closest approach to the Sun—and then retreats back into space again. Comets are at their brightest when nearest the Sun, where they shine by reflected sunlight; therefore, it seems logical to search for them in that region, even though their borrowed brilliance is greatly diminished by their bright sky background.

Dennis Anderson

Comets make fascinating sights in binoculars, as did Comet West in 1976.

Comets approach the Sun from every possible direction, and the planes of their orbits lie at every conceivable angle to the plane of the ecliptic. It is quite possible for a comet to approach the Sun from the side opposite Earth and remain invisible to us until its swift motion suddenly brings it into view from behind the Sun. The first comet I ever saw did exactly that.

On January 18, 1910, astronomers at Lick Observatory in California first saw this comet in broad daylight just east of the Sun. They estimated its brightness as exceeding that of Venus, which was then nearby in the sky. Several days and nights elapsed before I got to see it on the farm. In the meantime it had moved still farther to the east and by then I'm sure that it must have lost much of its brightness. It was the first comet of the year and so became known as Comet 1910a. I was only a youngster at the time and was not concerned with magnitudes and orbits, but I still vividly recall the ghostly appearance of this comet in our western evening sky. I remember well the sense of awe and wonderment it aroused in me.

The brightest sun-grazing comet on record, and another daylight one, was that of 1843. In 1668, 1880, 1882, and 1887 bright comets were observed that wheeled about the Sun at some 300 miles per second and were so near the solar surface that they must have passed right through its outer atmosphere. In more recent times, Comet Mrkos of 1957 and Comet Ikeya-Seki of 1965 were both sun-grazers; the latter was easily visible to the naked eye in full daylight if one merely blocked out the sunlight by hand. Comets also have been found from time to time on those occasions when our Moon cooperated by cutting off the sunlight with a total solar eclipse. The first of these was recorded by the Roman statesman Seneca in the first century. Others occurred in A.D. 418, 1860, 1871, 1882, 1893, 1948, and more recently in 1970. This latter was first detected with binoculars when only 12 degrees from the Sun.

An alert observer with shielded binoculars should detect any close Sun visitor more promptly and with more certainty than with the naked eye alone. Binoculars are also useful in estimating the magnitudes of comets and conducting a night-to-night watch for any changes in their physical appearance or any sudden flare-ups in their brightness. Abrupt changes sometimes do take place in comets, since they are rather frail affairs that are easily influenced by outside forces.

Estimating the brightness of a comet is a bit more complicated than judging the magnitude of a variable star. With the latter it is simply a matter of comparing the brightness of the variable with that of one or more stars of constant magnitude in the same field. However, since you won't find any convenient comparison comet to be pressed into service, you must make one from a star. To do this try to select a star of known brightness that, when thrown out of focus to match the apparent size of the comet, will equal it in magnitude. The designated brightness of the star you use represents the total, or integrated, brightness of the comet.

If you dissect a typical comet you will find that it consists of three main parts: the nucleus, the coma, and the tail. The nucleus is the starlike point that contains the solid material of the comet in a spongy, porous aggregation of dust, solidified gases, and ices of water, ammonia, methane, and carbon dioxide. The coma is the hazy envelope that surrounds the nucleus and is the one essential feature of every comet. It often is the only part that is visible. The coma is caused by the action of sunlight striking and driving off the gases and fine particles that compose the nucleus. A coma may be as big as a million miles in diameter.

The comet's tail is a continuation of the coma—the trailing out in space of the finer particles. It's nature's own bit of pollution, the particles left behind as the comet warms in the Sun's radiant energy. In this way the material in a comet's tail spread into the inner solar system. It follows, then, that each time a comet passes near the Sun it leaves some of its tail behind and becomes a little smaller and less massive. This loss is seldom a perceptible one, however, for even though the tail may be 100 million miles in length, its density is lower than that of smoke. Stars shine through a comet's tail with undiminished brightness.

As a general rule, a comet has no tail when first discovered in the mid-distant reaches of the solar system. At that point it appears as a hazy bit of light, not unlike a star seen through a mist. Only as the comet approaches the Sun, usually inside the orbit of the planet Mars, is the radiation pressure of sunlight and the solar wind sufficiently great to form the tail. You can see this solar wind by its effect: Like a giant fan mounted on the Sun it always blows the comet's tail away from it, even as the comet sweeps around the Sun and, tail foremost, heads off into space.

Many comets move around the Sun in elliptical or closed orbits that after a period of time bring them back near the Sun. Many of these periodic comets have been observed at successive returns. Most are classified as short-period comets. Perhaps the most frequently observed of these is Encke's Comet, which has a period of only 3.3 years. This comet has often attained naked-eye brilliancy and should, when well placed in the sky, be an easy object for binoculars.

Among the slowpokes of the skies are the long-period comets. This class includes many of the all-time greats: comets seen at midday close beside the Sun, awesome apparitions that creep across our skies but one time in a century or even in a thousand years. In this class, too, is the only comet that, if you are opportunely born and blessed with some longevity, you may see twice within your lifetime. This is Halley's Comet, the most famous of them all.

This historic comet has been seen at intervals of about 76 years ever since its first recorded appearance in 240 B.C. It was the first known periodic comet and the only one that has been bright enough at each return to be easily visible to the naked eye. Unlike most comets, this one was not named for its discoverer—who may well have been some nameless early human—but for the English scientist and Astronomer Royal, Edmond Halley, who discovered its periodicity and predicted its return. Halley watched this comet in 1682 and in plotting its orbit noticed in old records that similar bright comets had appeared in 1607 and 1531. As a period of about 76 years separated each of these three sightings, he suggested that the same comet was responsible. He then predicted that it would return in 1759. His forecast proved correct. Halley's Comet returned in 1835 and again in 1910, a sight that remains one of my own treasured childhood memories. The most recent return came in 1986.

Halley's Comet has the longest period of any comet seen making more than one trip around the Sun. Its orbit is a long, slender ellipse that extends well beyond the orbit of Neptune at aphelion, its greatest distance from the Sun, and makes its perihelion passage between the orbits of Mercury and Venus. Thus it crosses the orbits of eight of the planets, though because of its inclined orbit it makes no close approach to any of them. Unlike the planets it moves around the Sun in retrograde motion—from east to west.

In 1948 Halley's Comet rounded the outermost loop of its orbit and in 1968 it recrossed Neptune's orbit. Slowly accelerating, it passed above the orbit of Uranus in 1979. By 1982, when between the orbits of Saturn and Jupiter, the comet was found at about magnitude 24 on photographs exposed for it, though it then showed but a fuzzy, starlike image devoid of any tail.

Backyard astronomers got their first look at Halley's Comet in October 1985 as it brightened to 8th magnitude. The comet appeared low on the horizon for observers in the Northern Hemisphere, who had to travel south to such places as Australia and South America to glimpse the comet when it was at its brightest, between March and April 1986. But the disappointment of Halley's relatively poor performance was more than made up for by the first-ever views of a comet's nucleus provided by European and Soviet comet probes in March. From a distance of only a few hundred miles, the comet looked like a black, potato-shaped rock spouting bright jets of dust and gas.

As we await the next coming of Halley's Comet and its hoped-for extravaganza, let's keep our eyes and binoculars on the alert for the unexpected, which so often happens in the skies. Already, in this latter half of the twentieth century, we have met a number of interesting visitors, none of which had been predicted in advance. Early in the spring of 1957 the northwestern evening sky played host to naked-eye Comet Arend-Roland with its amazing spikelike tail that pointed toward the Sun. In August of that same year, the western skies were again enlivened, this time by 3rd magnitude, tail-switching Comet Mrkos. In October 1965 Comet Ikeya-Seki cut its capers in broad daylight close beside the Sun before it followed its long glowing tail back into space.

While searching the southern skies from his home in South Africa, amateur John C. Bennett came upon a small, diffuse, tailless comet not far from the bright star Achernar in southern Eridanus. Though only 8th magnitude at the time of discovery, this comet soon showed signs of a brighter future as it moved northward toward its perihelion, which it reached on March 20, 1970. At that time it was estimated at zero magnitude, making the comet's total brightness nearly equal to that of Vega, then showing in the east. As I saw it in early April with its 15-degree-long divided tail spread out against the dark pre-dawn sky, I thought it easily the finest comet I had seen since 1910.

Scarcely had Comet Bennett slipped from the grasp of the largest telescopes when still another bright voyager was glowing in our eastern morning skies. This one, too, came from the Southern Hemisphere, where in November 1975 a 17th-magnitude comet was found on photographic plates exposed three months before at the European Southern Observatory in Chile by astronomer Richard West. Like its bright predecessor, Comet West was headed for its brief reunion with the Sun when found; it arrived at perihelion on February 25, 1976. As it swung around the Sun it was estimated to be fully as bright as the planet Mars at a favorable opposition and was visible to the naked eye in daylight. The comet appeared at its best during the first week of March, when it was at 1st magnitude against a darker predawn sky. In my 7x35 binoculars, the tail was a streamered fan of marvelous complexity that curved to the north and west for some 25 degrees.

Comet West drew the largest audience of any comet in history, for it was well seen both north and south of the equator and all around the world. Its bright performance did much to atone for the dismal showing of stage-shy Comet Kohoutek three years before. As its finale, Comet West put on an act that even got an ovation from the astrophysicists: though it had come on stage with but a single head, it made its exit with no less than four. This quick-change act has been diagnosed as an early symptom of eventual disintegration, a penalty, perhaps, of a too-close approach during its command performance before the mighty Sun.

All who watched this splendid comet on those nights in early spring should treasure their remembrance of the occasion and rejoice at their good fortune to be in the audience of this one-time-only show. No Two-Timers Club will be formed to honor Comet West, for its period has been reckoned at around one million years.

The most spectacular cometary event occurred in July 1994, when the fragmented comet Shoemaker-Levy 9 slammed into the southern hemisphere of Jupiter. The events left blue-black bruises in the planet's cloud-tops that could be seen for weeks in a small telescope. Unprecedented press coverage turned the event into a circus-like atmosphere. The discovery that comets can break up and hit planets has already led to finding other examples of fragmented comets and the crashes left by them on other planets of the solar system.

16
Meteors

There are three distinct terms applied to these streaking voyagers, which I shall use here in the interests of accuracy, even though they all refer to the same object as it's seen in three different environments. First of all there is the term meteoroid, which applies to the object while it is still moving in outer space and invisible to us. If, however, this meteoroid enters the Earth's atmosphere, the resulting friction turns it into a glowing streak of light that now is called a meteor. In this stage it indeed is "something of the air." Finally, if the meteor survives the white heat of its passage through our atmosphere, it falls to Earth and becomes a meteorite, a suddenly arrested visitor from outer space.

In the study of meteors, accurate tracking of their visible path across the sky is of great importance, for only in this way can their individual characteristics and family traits be learned. By estimating the brightness of these streaks, the size of a typical meteor has been found to be about that of a grain of sand or, at most, a small pebble. By simultaneously observing identical meteors from two stations a few miles apart, it has been determined by triangulation that meteors become visible when about 60 miles above us and cease to glow when about 40 miles high. By determining the height of meteors, the length of their trails, and the interval of their visibility, the average velocity of meteors has been found to be about 25 miles per second.

The hours from midnight to dawn are the best for meteor observing, though for the observer they also are often the coldest and most inconvenient. During these hours the sky watcher is figuratively standing in the bow of the good ship Earth as it plows the sea of space. Thus he can clearly see all the flotsam the ship encounters. On the other hand, in the period between dusk and midnight, the watch is from the ship's stern, where the massive bulk of the ship itself blocks out everything but the slower

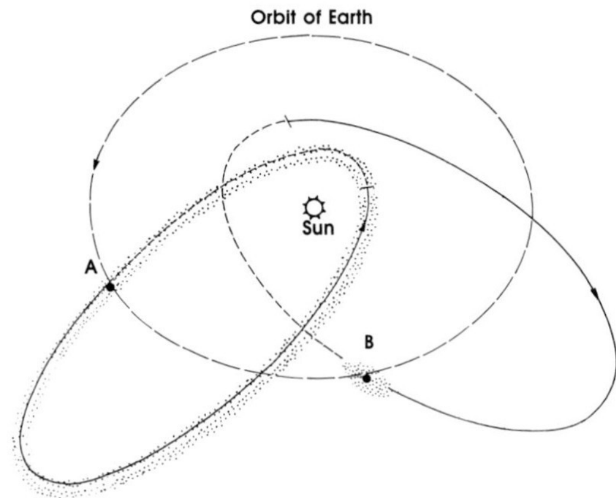

A: Earth passing through annual stream of meteoroids such as those of the Perseid shower.

B: Earth passing through swarm of meteoroids such as those of the Draconid shower.

The cross lines mark where the meteoroid orbits intersect the plane of Earth's orbit.

movements of the ship's wake. My drawing on page 73 shows why this is so.

Earth's two motions—its 18½-mile-per-second orbital movement around the Sun and its daily rotational turning on its axis—are both in the same direction, from west to east, and thus, in the after-midnight hours, the observer is moving into and meeting head-on any meteors in the Earth's path. This causes the after-midnight meteors to be swifter, brighter, and more numerous than those of the evening hours because Earth's speed is added to the normal velocity of the meteor. The evening meteors are those that are catching up with the moving Earth, so the Earth's motion must be subtracted from that of the meteor, leaving it with a velocity of only about 7 miles per second.

Here is a case where haste makes waste. The slower-moving evening meteor has the better chance of surviving the torrid passage through our atmosphere to land on Earth. Here, hopefully, it may be found and can spend its leisure years as a neatly labeled meteorite in some museum case where all can see and wonder at the awesome secrets it still withholds from us.

On any night of the year, if you are watching in a clear dark sky, an average of about six meteors per hour can be seen all about the sky, moving in various directions. On rare occasions, perhaps a dozen times each year, you will be greeted by what appears to be an unusual number of meteors that seem to be streaking outward in all directions from a definite location in the sky. You are watching a meteor shower. These showers occur on certain favored nights when Earth cuts through the orbit of a meteor stream, which often may flow in a broad elliptical orbit all the way about the Sun. In my table on page 125, I've listed the most important of these showers together with the date of maximum activity and the number of meteors per hour that you may expect at an average occurrence.

As an example, every year around the 12th of August Earth moves through the orbit of a meteor stream. If the night is clear and there is no

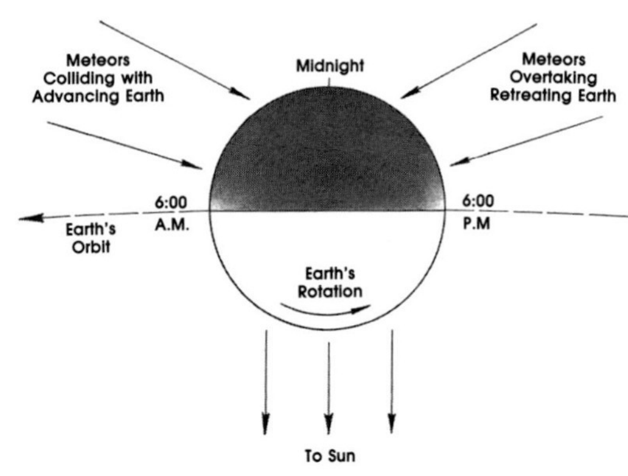

bright Moon in the sky, you are almost certain to see an unusual number of yellowish meteors streaking from the northeastern sky. If you face in that direction you may see as many as fifty of these per hour. This is the Perseid shower, so named because the streaks left by these meteors appear to originate at a point in the constellation Perseus not far from the famous Double Cluster. From that point, the streaks seem to radiate outward in all directions like the spokes of a wheel from its hub. Actually these meteors are all moving parallel to each other, but because of the illusion of perspective, they seem to be coming from one point— called the radiant—just as the parallel rails of a railway seem to come together in the far distance.

The Perseids form a steady and dependable shower with about the same hourly rate every year. From this we know that the individual meteoroids are evenly distributed throughout the entire stream. For this reason, the Perseids are an annual shower. In other showers with more eccentric habits the meteoroids are mostly concentrated in one vast swarm somewhere along the stream's long course. These are the periodic showers; only at rare intervals does Earth pass through one of these swarms and treat the fortunate observer to what must be the grandest of all sky shows. One such periodic shower, and the most noted of them all, is the Leonid shower, which on a number of occasions in the past has been seen to pour like a fiery fountain of stars from the Sickle in the constellation Leo.

The first Leonid shower to attract wide attention in America began on the evening of November 12, 1833. In the early morning hours at the height of the display it was estimated at as many as 35,000 meteors per hour were streaking across the sky in a shower that had now become a deluge. Here was clearly an instance where Earth had plunged headlong into a dense swarm of meteoroids. Further investigation showed that this swarm was in an orbit that wheeled around the Sun and then extended out in space to a point somewhat beyond the orbit of the planet Uranus in a period of 33 years. Oddly enough, this course proved to be an almost exact duplicate of the orbit of the Tempel-Tuttle Comet of 1866, which also comes to perihelion every 33 years. Another brilliant meteor shower came in 1866, but 33 years later, in 1899, the expected bright rerun was a disappointment, for the meteor swarm in the meantime had passed too close to both Saturn and Jupiter and had been drawn slightly from its course. At the next time around, in 1932, the meteors appeared only at a rate of one per minute, and in most areas the Leonids were written off as a shooting star attraction.

But to the amazement and delight of all, in 1966 the meteor swarm was back on course again. In the western United States at maximum in the morning hours of November 17, meteors were falling far too fast for counting. The display was rated as equal to the all-time spectacular of 1833, with estimates of 60,000 to 150,000 meteors per hour. Unfortunately, in many eastern states on this occasion rain triumphed over the shower, and even where clear skies prevailed the peak of the display came after sunrise. The next return of the Leonid swarm may happen in November 1999, if all interfering planets cooperate and keep their distance.

The fact that the orbit of the Leonid swarm is identical to that of a comet is not just coincidence. It is, rather, a confirmation of the close relationship between comets and meteors. Meteoroid swarms are the debris of old comets that have partially or completely disintegrated. Most of the showers we see have now been attributed to some known comet whose gradual deterioration has produced the stream of meteoroids that the Earth, in passing, may ignite into a fiery rain of meteors.

At the same time that you are watching the various shower meteors and tracing them to their radiant, you are certain to see other meteors that clearly do not belong to the shower in progress, for they don't share a common radiant. These are known as sporadic meteors: mavericks that have not, as yet, been observed as belonging to any known shower.

Binoculars have an important role in meteor observing, particularly in the recording of telescopic meteors, a term that, in a general way, refers to all meteors too faint to be caught with the naked eye. Here I shall designate as binocular meteors all those of 6th, 7th, and 8th magnitude, for these are well within the range of 7x binoculars.

Although low-power binoculars easily bring you at least ten times the total number of fixed stars that you can see with the naked eye, this ratio doesn't hold true for meteors. The broad expanse of sky dome that your unaided eye sees streaked with 50 meteors per hour on an August night dwindles to a Dipper bowl of sky when seen in your binoculars. For many years, I recorded all the telescopic meteors I saw while comet-hunting and observing variable stars with a telescope that had a field of view of 1½ degrees. My best annual total was 160 meteors, of which 18 were seen on a single busy night.

In observing binocular meteors you should select an area well up in the sky in order to avoid atmospheric contamination. Since estimating magnitudes is important here, the field can be that of a variable star that has a good sequence of comparison stars between 5th and 9th magnitudes, such

Important Meteor Showers

Name of Shower	Date of Maximum	No. Per Hour	Parent Comet
Quadrantids	Jan. 3	85	Not known
Lyrids	Apr. 22	15	Thatcher
Eta Aquarids	May 5	30	Halley
Delta Aquarids	July 28	20	Not known
Perseids	Aug. 12	100	Swift-Tuttle
Draconids *	Oct. 8	100	Giacobini-Zinner
Orionids	Oct. 22	20	Halley
Taurids	Nov. 3	15	Encke
Leonids*	Nov. 17	12	Tempel-Tuttle
Geminids	Dec. 14	95	3200 Phaethon?

*Denotes a periodic shower. All others are annual showers.

Notes on Showers

Quadrantids—The radiant of this annual shower lies in the now obsolete asterism Quadrans Muralis in northern Bootes. As it's low in the sky in early January, this display is best seen in the morning hours of January 3.

Eta Aquarids—This shower leisurely drips from the Water Jar, which rises shortly before dawn. Best seen from southern locations.

Perseids—Occasional members of this faithful family may appear from mid-July until late August. These are swift, bright, streak-leaving meteors that are best watched after midnight when the radiant is high in the sky.

Draconids—Often called Giacobinids from its parent comet discovered in 1900. I watched the finest display of this periodic shower on the evening of October 9, 1946, when for a time more than 100 meteors per minute were streaking across the sky from their radiant in the Head of Draco. Some of these were slow-moving fireballs brighter than the planet Venus. For me, that shower was a sky-wide exhibition of celestial fireworks.

Leonids—A few Leonids may be seen every year in the early morning hours around mid-November. They increase considerably in the years immediately preceding and following the great showers of the 33-year periods. This indicates that some meteoroids are scattered throughout the entire stream and that the main swarm itself is somewhat elongated in form.

as the variable R Leonis. Tracings or photocopies of such a chart masked to the diameter of your binocular field of view can be used as charts on which to draw a pencil line to indicate the streak made by each meteor you see. The direction of motion is indicated by an arrow attached to the line, which is also numbered to identify the streak for any notes you may make concerning magnitude, color, velocity, and length of streak. For this last, you should know the diameter, in degrees, of your field of view.

If a considerable number of charted meteors are moving in the same direction it's quite possible that they are shower meteors, in which case you should try to locate the radiant, for this is the main objective in observing telescopic meteors. To do this, select a second field at right angles to the streaks of the first field and some distance away. If you get a second series of parallel lines here, their intersection with the first series should lead you to the radiant.

At the time of a meteor shower of considerable activity you'll find it interesting to locate and explore the precise radiant point with your binoculars. As you approach the radiant you'll note that the meteor trails become shorter and shorter until, at the exact center of the radiant, some show no lateral motion at all but appear as brightening starlike points, for now they are moving directly toward you.

Meteor observing provides an interesting occupation for your binoculars. It can be of scientific value, too, for there is much to learn and the field is uncrowded. If you feel the fascination of these sudden streaks across the skies and would like to delve still further into the secrets revealed in the last brief moments of these frail voyagers from outer space, apply for more complete information to:

The American Meteor Society
Department of Physics and Astronomy
SUNY–Geneseo
Geneseo, NY 14454

17 Endless Trails

In this year-long parade of celestial sights, you may have found some favorites you and your binoculars have returned to again and again because of some special interest which they aroused in you. It may well have been the vagaries of some variable star, an unpredicted drop in the light of R Coronae Borealis, or the transient glow of ancient Mira in a barren autumn field of Cetus. Or perhaps it was a group of sunspots that formed or faded right before your eyes as they crept across our day-star's filtered face.

Perhaps the challenge of some less-traveled trail has appealed to you with its positive assurance that right now out in space an undiscovered comet is headed toward its rendezvous with the Sun. But no matter what star-lighted bypath intrigues you and turns your binoculars skyward, it will be an endless trail you follow. Each path winds through fields of wonder and enchantment that will surely lure you on to new challenges and new fulfillments through all the seasons of the years to come.

Appendix I

RECOMMENDED RESOURCES

Many beginning skygazers assume there exists somewhere the perfect book, software application, CD-ROM, or video that contains all the answers to every astronomy question they may have. If only this were true! Unfortunately, the cold fact is that no single information source covers all of astronomy, and the more questions you have, the more information resources you're going to need.

The main way to learn of new books and software are advertisements and review columns of magazines such as *Astronomy*. These typically review and describe newly published material of all kinds and are useful for giving you titles to look for.

Every inquisitive backyard astronomer should get to know the public library. Find the largest one you can get to conveniently and start to learn their holdings inside out. It's almost impossible to overstate the importance of this. Most backyard astronomers are largely self-taught and the process usually begins in the library. If you find yourself baffled by the catalog or the intricacies of the Dewey Decimal Classification System (astronomy is catalogued in the 520s), don't hesitate to ask at the reference desk for help.

Bookstores that stock new titles in astronomy range from terrible to outstanding. Superstores like Borders and Barnes & Noble often have many titles; independent bookstores often carry fewer titles, but they may be well-selected. Really, everything depends on the local demand for the subject and how well the store's manager does his or her job. If you don't see what you want, ask. A bookstore can order any item that's in print—though you may have to wait a few weeks for it to arrive.

Another source to explore is second-hand bookstores, although finding one with a good selection of astronomy titles is rare. With software and CD-ROM publications, the situation is much the same, only it's a computer store you're dealing with. Again, making your wants known to the manager will help.

So much for "information outlets." Now, what should you look for? To get an overview of the science of astronomy (as opposed to the practicalities of backyard observing), the best guide is one of the many "astronomy for poets" textbooks. There are dozens on the market and new editions appear every year. Ordinary bookstores don't carry them; check a college textbook store for one with a recent copyright date. For research results so new they haven't yet appeared in textbooks, regularly check the library for magazines such as *Science News* (a weekly) and *Astronomy* (a monthly).

As you get more involved in backyard observing, you may want to buy a telescope. An excellent place to start is the Observer's Guide, published annually by *Astronomy.* It contains full listings of all the equipment on the market (plus accessories, binoculars, and the like). It also tells beginners how to find the type of telescope that's just right for their needs.

A good guide to what your telescope can show you is *The Universe From Your Backyard,* by David Eicher (Kalmbach, 1988). Antonín Rükl's *Atlas of the Moon* (Kalmbach, 1990) and *The Moon* by Michael Kitt (Kalmbach, 1992) are essential items for any lunar observer.

Every backyard astronomer needs a star atlas, and Wil Tirion's *Sky Atlas 2000.0* (Cambridge University Press, 1981) has become a standard fixture in most observers' libraries. Although it isn't essential, you can also benefit from the two-volume catalogue that accompanies it, *Sky Catalogue 2000.0,* edited by Alan Hirschfeld and Roger W. Sinnott (Cambridge University Press, 1982 and 1985). The first volume contains basic data (names, positions, brightnesses, spectral types, distances, etc.) for some 45,000 stars down to magnitude 8.0. The second volume provides similar data for 21,000 double stars, variables, and non-stellar objects such as clusters, nebulae, and galaxies.

More advanced observers often choose *Uranometria 2000.0,* by Wil Tirion, Barry Rappaport, and George Lovi (Willmann-Bell, 1987 and 1988). Volume 1 covers the northern sky down to -6° declination, volume 2 maps the southern sky up to +6° declination. *A Deep Sky Field Guide* by Murray Cragin, James Lucyk, and Barry Rappaport (Willmann-Bell, 1993) tabulates, chart by chart, information on positions, brightnesses, sizes, etc., for all the non-stellar objects in the Uranometria.

For software, the choices are many. Expert Astronomer (IBM and Macintosh) is an inexpensive and basic "planetarium" program. It shows you the sky (using 9,100 stars) from any location at any given date and time. More advanced in sophistication (and price) is Dance of the Planets (IBM), an impressive tour de force that has superb graphics and the ability to help you visualize the solar system from anywhere within it, on any date. Voyager II (Macintosh) is a slick and powerful planetarium program with over 50,000 stars and 4,200 deep-sky objects. The world of astronomy software is large and growing rapidly; these are just the iceberg's tip. Check the issues of the astronomy magazines for updates and reports of new software if you want to stay current.

—Robert Burnham
Astronomy magazine

Appendix II

A PRONUNCIATION DICTIONARY

Achernar (AAK-her-nar)
albedo (al-BEE-do)
Albireo (al-BEER-e-o)
Aldebaran (al-DEB-a-ran)
Alpha Centauri (AL-fa sen-TAW-ree)
Alphecca (al-FEK-a)
Alpheratz (al-FER-atz)
Andromeda (an-DROM-e-da)
Antares (an-TAIR-eez)
aphelion (ap-HEEL-yon)
Aquarius (ah-KWAIR-ee-us)
Aquila (AK-wil-la)
Arcturus (arc-TUR-us)
Aries (AIR-eez)
Auriga (aw-REE-ga)
Aurigae (aw-REE-gay)
beta (BAY-ta)
Betelgeuse (BET-ul-jooz)
binary (BYE-nair-ree)
Bootes (bo-OOH-teez)
Camelopardalis (camel-o-PARD-al-is)
Canes Venatici (CAN-is ve-NAT-i-see)
Canis Major (CAN-is MAY-jor)
Canopus (can-NO-pus)
Capella (ka-PELL-a)
Carina (ka-REE-na)
Cassiopeia (kas-i-oh-PEE-ya)
Cepheid (SEFF-ee-id)
Cepheus (SEE-fee-yus)
Cetus (SEE-tus)
Chi Cygni (ky SIG-nee)
Coma Berenices (KOH-ma BARE-a-NEE-seez)
Cygnus (SIG-nus)
Deimos (DY-mose)
Delphinus (del-FEE-nus)
Deneb (DEN-eb)
Denebola (den-EB-o-la)
Draco (DRAY-ko)
Equuleus (e-KWOO-lee-us)
Eridanus (eer-i-DAN-us)
eta (AY-ta)
Europa (yur-RO-pa)
Fabricius (fab-RISH-yus)
Fomalhaut (FOE-mel-low)
galaxy (GAL-ax-ee)
Galileo (gal-a-LAY-o)
Gemini (JEM-in-nye)
Halley (HAL-lee)
Hevelius (he-VEL-ius)
Hyades (HY-a-deez)
Io (EYE-o)
Kochab (KO-chab)
Lacerta (la-SUR-ta)
Lepus (LEE-pus)
Leverrier (le-VARE-e-ay)
Libra (LEE-bra)
Lupus (LOO-pus)
Magellanic (majel-AN-ik)
mare (MAH-rey)
Messier (mess-ee-AY)
Mira (MY-ra)
Mizar (MY-zar)
Monoceros (mo-NOSS-er-os)
nebula (NEB-yoo-la)
nebulae (NEB-yoo-lee)
novae (NO-vay)
Ophiuchus (of-e-YOO-kus)
Orion (o-RYE-un)

Orionid (o-RYE-on-id)
Pegasus (PEG-a-sus)
Perseus (PER-see-us)
Piazzi (pee-AHT-zee)
Pisces (PIS-eez)
Piscis Austrinus (PIS-is aw-STRY-nus)
Pleiades (PLEE-ya-deez)
Polaris (po-LAIR-us)
Praesepe (pray-SEPH-ee)
Procyon (PRO-see-on)
Ptolemy (TOL-e-mee)
Regulus (REG-yoo-lus)
Rigel (RYEjell)
Roemer (RAY-mer)
Sagitta (sag-IT-a)
Sagittarius (saj-i-TAIR-ee-us)
Scutum (SCOOT-um)
Sirius (SEER-ee-us)
Spica (SPY-ka)
Taurus (TAW-rus)
theta (THAY-ta)
Tycho Brahe (TY-ko BRA-hee)
Uranus (YOOR-a-nus, yoo-RA-nus)
Vega (VAY-ga)
Vulpecula (vul-PEK-yoo-la)
zodiac (ZO-dee-ak)
zodiacal (zo-DYE-a-kal)

Appendix III

THE BRIGHTEST STARS

The twenty-one stars on page 133 are listed in order of decreasing brightness. The stars marked with an asterisk cannot be seen from mid-northern latitudes, although Canopus can be glimpsed briefly in winter from our most southern states. Also listed is the constellation to which each star belongs as well as a space for you to record the date when you first identified the star.

Luke Dodd

The Orion Nebula is the finest cloud of glowing gas visible from the Northern Hemisphere.

Star	**Constellation**	**Magnitude**	**Date**
Sirius	Canis Major	-1.46	
*Canopus	Carina	-0.72	
Arcturus	Bootes	-0.04	
*Alpha Centauri	Centaurus	0.00	
Vega	Lyra	0.03	
Capella	Auriga	0.08	
Rigel	Orion	0.12	
Procyon	Canis Minor	0.38	
Betelgeuse	Orion	0.4 to 1.3	
*Achernar	Eridanus	0.46	
*Beta Centauri	Centaurus	0.61	
Altair	Aquila	0.77	
Aldebaran	Taurus	0.85	
Antares	Scorpius	0.96	
Spica	Virgo	0.98	
Pollux	Gemini	1.14	
Fomalhaut	Piscis Austrinus	1.16	
Deneb	Cygnus	1.25	
Regulus	Leo	1.35	
Alpha Crucis	Crux	1.41	
*Beta Crucis	Crux	1.88	

Appendix IV

THE CONSTELLATIONS

Constellation	Abbreviation	Date of First Sighting
Andromeda	And	
Antlia	Ant	
Apus	Aps	
Aquarius	Aqr	
Aquila	Aql	
Ara	Ara	
Aries	Ari	
Auriga	Aur	
Bootes	Boo	
Caelum	Cae	
Camelopardalis	Cam	
Cancer	Cnc	
Canes Venatici	CVn	
Canis Major	CMa	
Canis Minor	CMi	
Capricornus	Cap	
Carina	Car	
Cassiopeia	Cas	
Centaurus	Cen	
Cepheus	Cep	
Cetus	Cet	
Chamaeleon	Cha	
Circinus	Cir	
Columba	Col	
Coma Berenices	Com	
Corona Australis	CrA	
Corona Borealis	CrB	
Corvus	Crv	
Crater	Crt	
Crux	Cru	

Constellation	Abbreviation	Date of First Sighting
Cygnus	Cyg	
Delphinus	Del	
Dorado	Dor	
Draco	Dra	
Equuleus	Equ	
Eridanus	Eri	
Fornax	For	
Gemini	Gem	
Grus	Gru	
Hercules	Ger	
Horologium	Hor	
Hydra	Hya	
Hydrus	Hyi	
Indus	Ind	
Lacerta	Lac	
Leo	Leo	
Leo Minor	LMi	
Lepus	Lep	
Libra	Lib	
Lupus	Lup	
Lynx	Lyn	
Lyra	Lyr	
Mensa	Men	
Microscopium	Mic	
Monoceros	Mon	
Musca	Mus	
Norma	Nor	
Octans	Oct	
Ophiuchus	Oph	
Orion	Ori	
Pavo	Pav	
Pegasus	Peg	
Perseus	Per	
Phoenix	Phe	
Pictor	Pic	

Constellation	Abbreviation	Date of First Sighting
Pisces	Psc	
Piscis Austrinus	PsA	
Puppis	Pup	
Pyxis	Pyx	
Reticulum	Ret	
Sagitta	Sge	
Sagittarius	Sgr	
Scorpius	Sco	
Sculptor	Scl	
Scutum	Sct	
Serpens	Ser	
Sextans	Sex	
Taurus	Tau	
Telescopium	Tel	
Triangulum	Tri	
Triangulum Australe	TrA	
Tucan	Tuc	
Ursa Major	UMa	
Ursa Minor	UMi	
Vela	Vel	
Virgo	Vir	
Volans	Vol	
Vulpecula	Vul	

Appendix V

LIFE-LIST OF CELESTIAL OBJECTS

This list includes all the named features mentioned in the pages of this book that can be observed with low-power binoculars. Blank spaces have also been provided at the end of the list so you can record names, dates, and comments relating to special events and sightings—such as eclipses, comets, meteors, novae, occultations, etc.—which cannot otherwise be classified here.

Object	Date	Comments
STARS		
Polaris		
The Pointers		
Alkaid		
Kochab		
Mintaka		
Orion's Belt		
Orion's Sword		

Object	**Date**	**Comments**
Zeta Tauri		
Pi Puppis		
Castor		
Thuban		
Gamma Cephei		
Alpha Cephei		
The Sickle		
Gamma Leonis		
Zeta Leonis		
Epsilon Leonis		
Denebola		
Tau Leonis		

Object	Date	Comments
Alphard		
Cor Caroli		
Alphecca		
Mu Cephei		
Beta Serpentis		
The Keystone		
Zeta Scorpii		
Beta Librae		
The Water Jar		
DOUBLE STARS		
Mizar		
Alcor		

Object	Date	Comments
Gamma Leporis		
Zeta Geminorum		
Delta Corvi		
Nu Draconis		
Delta Cephei		
Theta Serpentis		
Kappa Herculis		
Mu Scorpii		
Nu Scorpii		
Alpha Librae		
Epsilon Lyrae		

Object	Date	Comments
Zeta Lyrae		
Albireo		
61 Cygni		
Omicron Cygni		
Alpha Capricorni		
Beta Capricorni		
Omicron Capricorni		
Alpha Cassiopeiae		
Beta Piscium		
66 Ceti		
Alpha Ceti		

Object	Date	Comments
NEBULAE AND CLUSTERS		
M42 (The Orion Nebula)		
The Hyades		
Ml		
M45 (The Pleiades)		
M79		
M41		
NGC 2244		
M50		
M55		
M44 (The Beehive)		
M67		

Object	Date	Comments
M49		
M58		
M60		
M87		
M53		
M64		
M51 (Whirlpool Galaxy)		
M3		
M5		
M13 (Hercules Cluster)		
M92		

Object	Date	Comments
M80		
M4		
M6		
M7		
M8 (Lagoon Nebula)		
M20 (Trifid Nebula)		
M17		
M22		
M24		
M11		
M71		

Object	Date	Comments
M27 (Dumbbell Nebula)		
M30		
M52		
NGC 663		
Double Cluster		
M34		
M15		
M2		
NGC 7009 (Saturn Nebula)		
M77		
M31 (Andromeda Galaxy)		

Object	Date	Comments
M33		
M36		
M37		
M38		
VARIABLE STARS		
Betelgeuse (Alpha Orionis)		
R Leporis		
Eta Geminorum		
Zeta Geminorum		
R Leonis		
R Hydrae		
R Coronae Borealis		

Object	Date	Comments
Mu Cephei		
Delta Cephei		
R Serpentis		
Alpha Herculis		
RY Sagittarii		
Beta Lyrae		
Chi Cygni		
Eta Aquilae		
R Scuti		
Gamma Cassiopeiae		
Algol (Beta Persei)		

Object	Date	Comments
Mira (Omicron Ceti)		
THE SUN		
Sunspots		
Umbra		
Penumbra		
Faculae		
THE PLANETS		
Mercury		
Transit		
Venus		
Phases		
Mars		
Retrograde motion		

Object	**Date**	**Comments**
Jupiter		
Io		
Europa		
Ganymede		
Callisto		
Saturn		
Titan		
Uranus		
Neptune		
THE MOON		
Earthshine		
Terminator		

Object	Date	Comments
1-day Moon		
Mare Crisium		
Mare Fecunditatis		
Mare Serenitatis		
Mare Tranquillitatis		
Mare Nectaris		
Mare Imbrium		
Mare Frigoris		
Mare Nubium		
Mare Humorum		
Langrenus		

Object	**Date**	**Comments**
Vendelinus		
Petavius		
Cleomedes		
Furnerius		
Atlas		
Hercules		
Endymion		
Macrobius		
Piccolomini		
Theophilus		
Cyrillus		

Object	Date	Comments
Catharina		
Posidonius		
Fracastorius		
Eudoxus		
Aristoteles		
Cassini		
Hipparchus		
Albategnius		
Manilius		
Aristillus		
Plato		

Object	**Date**	**Comments**
Archimedes		
Autolycus		
Ptolemaeus		
Alphonsus		
Arzachel		
Walter		
Maginus		
Tycho		
Clavius		
Eratosthenes		
Longomontanus		

Object	**Date**	**Comments**
Copernicus		
Bullialdus		
Aristarchus		
Kepler		
Grimaldi		
Oceanus Procellarum		
Cape Agarum		
Palus Somnii		
Alpine Valley		
The Alps		
The Apennines		

Object	Date	Comments
Lunar Rays		
ASTEROIDS		
Ceres		
Pallas		
Vesta		
METEOR SHOWERS		
Quadrantids		
Lyrids		
Eta Aquarids		
Delta Aquarids		
Perseids		
Draconids		

Object	Date	Comments
Sinus Iridium		
Orionids		
Taurids		
Leonids		
Geminids		

Index

R

S

T

U

V

W

Z